Pakistan's Arms Procurement and Military Buildup, 1979–99

Also by Ayesha Siddiqa-Agha

CURBING THE WEAPONS OF CIVILIAN DESTRUCTION

MANAGEMENT OF LIGHT WEAPONS PRODUCTION IN PUBLIC AND PRIVATE SECTORS: A View from Pakistan

MARITIME COOPERATION BETWEEN INDIA AND PAKISTAN: Building Confidence at Sea

PAKISTAN'S NUCLEAR DOCTRINE AND COMMAND AND CONTROL: Impact on Regional Stability

Pakistan's Arms Procurement and Military Buildup, 1979–99

In Search of a Policy

Ayesha Siddiqa-Agha

Foreword by Lawrence Freedman

Published by
PALGRAVE MACMILLAN
Houndmills, Basingstoke, Hampshire RG21 6XS and
175 Fifth Avenue, New York, N. Y. 10010
Companies and representatives throughout the world

PALGRAVE MACMILLAN is the global academic imprint of the Palgrave
Macmillan division of St. Martin's Press, LLC and of Palgrave Macmillan Ltd.
Macmillan® is a registered trademark in the United States, United Kingdom
and other countries. Palgrave is a registered trademark in the European
Union and other countries.

ISBN 978-0-333-73172-7

This book is printed on paper suitable for recycling and
made from fully managed and sustained forest sources.

A catalogue record for this book is available from the British Library.

Library of Congress Catalog Card Number: 00–065222

Transferred to digital printing 2005

To the memory of my parents and Kono

Contents

List of Tables

List of Figures and Map

Acknowledgements

It was not easy doing this study. I needed to acquaint myself with a variety of concepts and find information on the subject. I am extremely grateful to Lawrence Freedman, Professor Emeritus of the Department of War Studies, Kings College, London University, for helping me achieve my objective. With his immense academic capabilities and insight, he guided me through my work. I would also like to thank Dr Saadet Deger and Professor Somnath Sen; their directions were essential.

Initially I faced insuperable hurdles in collecting information. I could not have achieved my results without the help of many senior military officers, diplomats and academics. Their cooperation was a mark of their open-mindedness and genuine desire to ensure the completion of an academic work on the subject. I am particularly grateful to Lt General (Retd) Farrukh Khan, Maj. General Khalid Bashir, Maj. General Salimullah, Lt General Jamsheed Malik, Maj. General (Retd) Ahmed Ali, Maj. General (Retd) Talat Masood, Lt General (Retd) Hameed Gull, Air Chief Marshal (Retd) Abbas Khattak, Air Marshal Aliudin, Admiral Abdul Aziz Mirza, Vice Admiral (Retd) Obaidullah Khan, Dr Abdul Qadeer Khan, Begum Abida Hussain, Mr Niaz A. Niak, Dr Ian Anthony, and Dr Kenneth Martin. I am extremely thankful to them for reposing the confidence in me and for the support they provided. I must also mention the help given to me by the former Military Accountant General Mr Majeed Khan, and Controller Factories Accounts Mr Abid Salaam. I would like to extend my special thanks to Admiral (Retd) Fasih Bokhari who provided me with the rare opportunity of understanding the military environment. The list of people that I would like to acknowledge for their cooperation is long; I would like to extend my appreciation to all those people who provided me with the opportunity to talk to them about the subject of my research.

This acknowledgement would be incomplete without mentioning the help rendered to me by my husband Omar. Without his cooperation, support, patience, encouragement and final editing I would have not been able to complete my work. I also owe much to Dr Shahla Haeri, Dr Jameel Jalabi, Khurram Tajammul, Sandra Tharumalingam and Manoj Narsey. Their motivation was vital in my decision to carry out my research and complete the book. The help provided by Choti at home was also invaluable.

Preface

This book undertakes a comprehensive analysis of Pakistan's arms procurement and military decision-making from 1979 to 1999. The study did not focus on the single perspective, strategy. The objective was to explain the subject with all its intricacies. Indubitably, Pakistan's relations with its traditional adversary, India, had a significant impact on the way decisions were made in procuring equipment and building nuclear deterrence. However, the peculiar decision-making environment, the role played by the military in defense policy-making, the evolution of democracy during this period and its impact on relevant decision-making, and the national and industrial constraints are elements that had an equally forceful impact on Islamabad's military buildup.

Undertaking such a study was certainly not an easy task. The Pakistani military establishment, in particular, is known for its high sense of secrecy. The parliament, that is, technically speaking, the body that approves military spending, has much less information than it ought. Military modernization planning is one area in which there is little information available domestically and internationally. The core objective of this book is to reveal how these decisions are made and what factors have impact on this policy-making. The period in which this research started was particularly good: doors were relatively open for a researcher in this area. There were a number of areas where information was not available and it is hoped that this study will help with further research. Transparency of decisions, after all, is one of the primary features of a democratic process that is currently not a strength of the Pakistani political system. It is also hoped that with this study the research community, especially in Pakistan, will get an insight, which may help it to re-think and redefine national strategic priorities.

Ayesha Siddiqa-Agha
Islamabad, Pakistan

Foreword

How does a country that is both chronically insecure and chronically poor go about defending itself? That is essentially the question posed by Ayesha Siddiqa in this book. She describes with candor as well as sympathy the dilemma faced by Pakistan in working out how to cope with sources of threat and instability on both its eastern and northern borders, stressing the centrality of the long-standing tension with India. This turned into an arms race during the 1980s, in both the nuclear and conventional fields. The only way that Pakistan can begin to stay in touch with the much powerful India is by gaining the support of others, especially the United States. So long as the Russians were fighting in Afghanistan, this could be used by Islamabad to persuade Washington to keep it supplied with major weapons. The Russian withdrawal from Afghanistan together with the end of the Cold War has left Pakistan more exposed, with Washington unwilling to support an arms buildup directed against India and censorious when it comes to Pakistan's nuclear ambitions.

With little spare cash, Pakistan is disadvantaged in the international arms market and with only a modest technical base there are limits to what can be achieved on its own resources. None the less, in a country in which the military plays such a prominent political role, and can so ensure a favorable share of public expenditure, the effort is made to acquire weapons and sustain a serious combat capability. By penetrating deep into the structure of Pakistan's procurement decision-making, Dr Siddiqa-Agha reveals the system at work behind the organizational charts. She highlights the strength of the military, and the relative weakness of the civilian bureaucracy, in this process as well as the opportunities for corruption.

The succession of case studies demonstrate these factors at work as Pakistan experiences changes in both its strategic environment and domestic politics. With remarkable detail, Dr Siddiqa-Agha describes the fate of major acquisitions planned by each of the three services. Her conclusion is constructive though not optimistic. Pakistan needs to concentrate on a defensive capability *vis-à-vis* India while at the same time searching for a new understanding with Delhi. Meanwhile the civilian management of the defense sector needs strengthening –

although this probably depends on Pakistan's democratic system being strengthened first.

LAWRENCE FREEDMAN

Glossary

ACAS	Assistant Chief of Air Staff
ACNS	Assistant Chief of Naval Staff
AEW	Airborne Early Warning
AIP	Air Independent Propulsion
ARDE	Armament Research and Development Establishment
AoG	Aircraft on Ground
ASW	Anti-Submarine Warfare
AWACS	Airborne Warning and Control System
AWC	Air Weapons Complex
AWT	Army Welfare Trust
BCCI	Bank of Credit and Commerce International
BJP	Bharatiya Janata Party
BVR	Beyond Visual Range
CAR	Central Asian Republics
CCC	Commander-in-Chief's Committee
CETS	Contractor Engineering Technical Services
CBU	Completely-Built-Unit
CGS	Chief of General Staff
CKD	Completely-Knockdown-Kit
COMKAR	Commander Karachi
COMLOG	Commander Logistics
COMPAK	Commander Pakistan Fleet
CMLA	Chief Martial Law Administrator
CTBT	Comprehensive Test Ban Treaty
DCAS	Deputy Chief of Air Staff
DCC	Cabinet Committee for Defense
DCNS	Deputy Chief of Naval Staff
DESO	Defense Sales and Export Organization
DESTO	Defense Science and Technology Organization
DG DP	Director General (Defense Purchase)
DG MP	Director General (Munitions Production)
DISAM	Defense Institute of Security Assistance Management
EEZ	Exclusive Economic Zone
FMF	Foreign Military Financing
FMS	Foreign Military Sales
GHQ	General Headquarters

GOC	General Officer Commanding
GST	General Sales Tax
HIT	Heavy Industries, Taxila
IAF	Indian Air Force
IJI	*Islami Jamhoori Ittehad*
IMF	International Monetary Fund
IN	Indian Navy
IOP	Institute of Optronics
IPP	International Power Producers
IRBM	Intermediate Range Ballistic Missile
ISI	Inter-Services Intelligence
ISPR	Inter-Services Public Relations
JCSC	Joint Chiefs of Staff Committee
JKLF	Jammu and Kashmir Liberation Front
JSHQ	Joint Staffs Headquarters
KARF	Kamra Avionics and Radar Factory
KRL	Kahuta Research Laboratories
KSEW	Karachi Shipyard and Engineering Works
LoC	Line of Control
MAG	Military Accountant General
MBT	Main Battle Tank
ME	Margalla Electronics
MIRDC	Mechanical Industrial Research and Development Corporation
MoD	Ministry of Defense
MoF	Ministry of Finance
MoFA	Ministry of Foreign Affairs
MQM	*Mohajir Qaumi* Movement
MVRDE	Military Vehicles Research and Development Establishment
NCCA	Nuclear Command and Control Authority
NDC	National Development Complex
NHQ	Naval Headquarters
NPT	(Nuclear) Non-proliferation Treaty
NRDA	Naval Research and Development Authority
NSC	National Security Council
NWC	National Weapons Complex
NWFP	North-West Frontier Province
OEM	Original Equipment Manufacturer
PAC	Pakistan Aeronautical Complex
PACO	Pakistan Automobile Corporation

PADS	Pakistan Air Defense System
PAEC	Pakistan Atomic Energy Commission
PAF	Pakistan Air Force
PIA	Pakistan International Airlines
PLA	People's Liberation Army (China)
PN	Pakistan Navy
PNE	Peaceful Nuclear Explosion
POFs	Pakistan Ordnance Factories
PPP	Pakistan People's Party
RAW	Research and Analysis Wing (Indian intelligence agency)
SLOC	Sea Lanes of Communication
SSG	Special Services Group
SKD	Semi-Knockdown-Kit
TML	Trans-Mobile Limited
UAV	Unmanned Air Vehicle
WAPDA	Water and Power Development Authority

Introduction

Why countries engage in military conflict, hostile relations, arms buildup, and proliferation of weapons of mass destruction is a question that has attracted the attention of a number of analysts. Theories ranging from security concerns to the bureaucratic imperative have been expounded to explain this particular behavior. Such issues have become increasingly important, especially since the collapse of the Soviet Union, and the threat of increase in the proliferation of weapons of mass destruction. The literature, however, remains scant with the need for more case studies contributing to an understanding of the subject. Moreover, the general tendency in the literature is to look at defense decision-making from more than one angle. Arms procurement and nuclear proliferation need to be understood from a holistic decision-making standpoint, which has been the aim of this book. The purpose is to study Pakistan's arms procurement and military buildup decision-making. The idea was to carry out a comprehensive analysis that would be based on an examination of all those factors that contribute towards the policy-making process.

When I began work on this project, there were no such studies on Pakistan. The only other analysis on Pakistan's arms procurements, by Ian Anthony, was based on two aspects: strategic imperative and arms trade.[1] Anthony's work, in fact, was a case study on the trends of arms transfer to the Indian subcontinent describing the various categories of weapon systems supplied to Islamabad at different times. His main argument was on how a convergence of views in both the US and Pakistan regarding the communist threat to South Asia resulted in the latter obtaining arms.[2] From a domestic angle, Anthony basically used the traditional Richardson action–reaction model to explain Pakistan's arms procurement. According to his analysis, Islamabad has obtained

1

arms because of New Delhi.[3] He did mention the economic factors and military's role in decision-making,[4] but this was more of a passing reference with no details of how the policy-making mechanism worked. Anthony was not the only one to use such a limited framework. The action–reaction and foreign alignment model was used by Pervaiz Cheema in analysing Pakistan's arms procurement and general defense decision-making.[5] His main argument circulated around Pakistan's sense of threat from India and its alignment with the US that facilitated Islamabad in acquiring weapons wanted to counter a hostile New Delhi. Cheema added another angle to Pakistan's procurement debate which related to arms transfers and Pakistan's foreign alignments.

In fact, most studies on Pakistan's defense policy-making, arms procurement, nuclear decision-making, or defense spending were conducted in the 1980s and in the early 1990s.Their focus was on viewing the defense policy-making process from a conflict standpoint. For example, Deger and Sen used this methodology to examine Pakistan and India's defense expenditure. In their view, Pakistan's threat perception, which determined its defense expenditure, came solely from India.[6] Wirsing tried to bring a certain variation to the debate by adding the military technological factor.[7] In his view, the conventional weapons technology, later the acquisition of non-conventional defense technology by Pakistan, was based on the military technological disparity which Islamabad experienced *vis-à-vis* New Delhi. This was a perspective subscribed to by other authors as well, such as Ali, Anthony, Aronson, Cheema, Creveld, and Moshaver.[8] The only academic work that talked about personal interests of political leaders such as Zulfiqar Ali Bhutto and General Zia-ul-Haq relate to Kapur's book on Pakistan's nuclear program.[9] That book, however, predominantly presents an Indian view on the subject. This was in addition to the fact that the book focused entirely on the bureaucratic and personal imperative rather than knitting it with the strategic imperative and other elements. The limited literature on Pakistan's arms production also used the security concern framework to show why Islamabad decided to establish an indigenous weapons manufacturing industry.[10] I found Mathews' paper, in which he discussed Pakistan's defense production, different and original; nevertheless, it neither talked about the relevant decision-making.[11] In addition, the journalistic pieces one comes across on defense decision-making in Pakistan do not present an integrated approach and do not answer all the questions as to why Islamabad engaged in an arms race with India, procured particular types of weapon systems, or chose to develop a nuclear deterrent. One comes across similar shortcomings in the literature of a number of other developing countries.

For over fifty years after independence from the British rule, the standard explanation given for Pakistan's arms procurement and military buildup pertained to security imperative and Islamabad's concern for the threat posed by neighboring India. Again, it was India's hand seen behind internal insecurity that has plagued the country for almost the past twenty years. It must be remembered that during the 1960s, when government's misdirected policies resulted in political turmoil in the eastern wing, a similar argument was used. Yet, there were times when questionable decisions were taken. Moreover, the entire logic for insecurity, threat perception and high defense spending at the cost of socioeconomic growth had to be analysed. Was it really and purely the external threat, or were there other factors influencing government decisions, especially in the defense sector? It was vital to find some answers to these questions because external threat is the principal prominent framework that has been used to explain military buildup decisions in Pakistan and for other countries faced with a hostile neighbor. When I started working on the subject I was trapped within this traditional framework. However, a deeper analysis of a number of weapons procurement decisions made me wonder about the credibility of using the security imperative as the only explanation for the huge investment of resources that successive regimes in Pakistan have made in the defense sector. Elements such as inter-services rivalry, the influence of the political government *versus* the military establishment, the nature of domestic power politics affecting defense decision-making, and organizational and personal bias for particular programs or weapon systems were some of the factors that were almost totally excluded in earlier works.

These are some of the areas that one could find in the literature on defense procurement and defense decision-making in the developed world. Unfortunately, one could not see many examples of a schematic shift in the literature pertaining to developing countries. Until the end of the Cold War, most of the academic works on arms procurement in developing countries predominantly presented the arms transfers perspective. Researchers viewed arms acquisitions, not as an independent policy-making process taking place in the recipient states, but more in connection with what supplier states thought of them. They also looked into the strategic importance of the recipient states for suppliers that facilitated the transfers, and other angles that were based on suppliers' policies. Research on defense production was also not qualitatively very different. Having established the point that most of the Third World states were involved in indigenous weapons production owing to security, enhancement of national image, and to a certain extent, economic reasons, the focus would normally shift to technological issues. One can find interesting debate on the impact of the transfer of technology on international arms trade and other relevant issues. The schematic shift in the literature on the subject related to many of the developed countries was possible mainly as a result of better transparency of the policy-making process. This is, indeed, not a characteristic of the political and cultural environment of most developing countries, especially Pakistan. An academic work on arms procurement has never been undertaken before now because of unavailability of published sources and the military or government's resistance towards discussing such issues in the interests of public awareness within the country.

This work is now only possible because of General (Retd.) Mirza Aslam Baig, who became the Army chief after General Zia's death, changing the policy in favour of some openness about military affairs. Thus it was possible for me to discuss numerous issues with a large number of serving and retired military personnel. As an institution, I found the Army and Navy more open than the Air Force, with the large size of the Army reflecting the influence it enjoys in the country's power politics. As for the Navy, its relative openness is linked with the service's need to publicize its requirements. The PAF, on the other hand, has always had a closed culture: having the confidence that it is indispensable for national defense, the service continues to be secretive.

Indubitably, the task was not an easy one but, with access to published data by some of the international research institutions and cross-referencing the information I was provided with, it was possible for me to arrive at certain conclusions. This *perestroika*, nevertheless, was to be short-lived. Pakistan's military, unfortunately, is again almost as closed as it was at the end of the 1980s and is now accompanied by an attitude of 'denial' that does not permit the government machinery to look beyond security, external threat and secrecy. The new military government, aiming at ensuring better efficiency, progress and growth, must first shift the focus from external threat to the internal dynamics of security. This realignment of national priorities requires a major shift in the state's foreign policy planning and possible downsizing of the military establishment. Whether the armed forces would support such a course of action depends upon the military management's ability to sacrifice short-term organizational benefits for long-term national interests.

A review of the literature on arms trade, arms production and decision-making left me searching for answers as to how to analyse arms procurement decision-making in Pakistan. One could get some answers from the arms trade literature about why suppliers such as the US, France, Britain or China gave arms to Islamabad, and find some explanations about why Pakistan decided to start indigenous weapons production. My perception while doing the literature survey, however, was that the existing studies did not provide the necessary linkage between strategy, industrial capabilities, official objectives and decision-making. Added to this, was public opinion in Pakistan; talking to different people in the country I came across two sets of views. There were those who thought that arms procurement was all about financial kickbacks and others who believed that weapons acquisition decisions were neatly tied in with the strategic needs of the country.

I couldn't completely reconcile myself to either explanation. At times, generalizations help in understanding the moods of an individual, group or society but it is an unscientific way of presenting a hypothesis. Besides, if I used one of these standpoints, there were still many questions that remained unanswered. There was also a third group that viewed defense decision-making as a process totally controlled by the military. Although it is true that the armed forces are the dominant actor, I had to be careful in applying this perception during the entire period of the present study. From 1977–88 the Army definitely controlled policy-making in the country but, after General Zia died the country started its journey back to democracy. There were frequent interruptions to civil rule and the army's attempts to destabilize the system. In addition, inter-services rivalry affected decisions. Moreover, there was the impact created by the US arms embargo imposed on Islamabad in 1990 together with the changes in Beijing's attitude towards Pakistan. My idea was to try to construct a comprehensive picture of the decision-making process taking into account all elements that contributed towards the policy-making process and its final outcome.

From a decision-making standpoint, I found the period from 1979–99 interesting for analysis. This was for three reasons: first, there was a lot going on in the military-strategic front. At the beginning of 1980 Pakistan's official threat perception had intensified with Islamabad projecting a two-front situation. There was concern about Soviet intentions of reaching the 'warm-waters' by invading Pakistan. The main threat continued to be from India, however. In fact, policy-makers in Islamabad saw a connection between the Indian threat and the situation on Pakistan's northern border. The Soviet invasion of Afghanistan coincided with an increase in tension with India resulting in both governments' adopting a more aggressive stance towards each other.

The US arms transfer to Pakistan in the 1980s that was to help Islamabad counter the Soviet threat was disturbing for New Delhi. Islamabad's arms acquisitions were viewed as destabilizing the strategic balance planned by India for the South Asian region. This resulted in the enhancement of the arms race in the region. Driven by its strategic goals, India tried to offset any advantage to Islamabad by increasing India's arms purchases and generally increasing military technological competition. Policy-makers of both countries resorted to the acquisition of conventional and non-conventional military technologies to address the growing threat. This competition precipitated the bilateral tension resulting in certain developments during this period. Several events of military-strategic significance took place, which led to more tension. Not only that, the continual increase in hostilities was encouraged in order to create the logic for a military buildup. These happenings had a bearing on the arms procurement decision-making process in both India and Pakistan.

The second reason was that this was a period when arms procurement decisions were influenced by variations in the arms suppliers' policies. The most significant were the fluctuations in Pakistan–US weapons transfer ties. Providing Islamabad with military hardware was dependent upon the 'ebb and flow' of American interests in the South Asian region. From this perspective, this period can be divided into three phases: (a) the position in 1979, (b) 1980–88, and (c) post-1988. In each phase, Washington's arms transfer policy regarding Pakistan was different depending on the priorities of American policy-makers at the time. For instance, when containing the growing power of communism was the main concern in the US, other issues such as nuclear proliferation elsewhere were given less importance. Then the American views on nuclear proliferation did not hinder arms transfer to Islamabad the way it did before 1979 or after 1988. Similarly, Pakistan's military links with China were based on a different frequency than the period prior to 1979. Although Islamabad's bilateral ties with Beijing have been more stable than with Washington, the chemistry of this link altered as well.

Third: there were rapid developments in the domestic political arena that affected the manner in which decisions were made. It was the changes in the political scene during this period that led to the strengthening of actors other than the military who influenced decisions. A period of military rule was followed by re-introduction of democracy in 1988. Civil rule started in 1985 but it could not sufficiently anchor itself. However, this was a period when there were more instances of financial corruption in arms deals. There were cases where top civilian decision-makers approved weapon purchases for personal financial gain. The persistence of this was an inherent disincentive to the improvement of the decision-making apparatus at the middle and top management levels. The weaknesses within the system, inefficient bureaucratic control and lack of knowledge of weapon systems in the MoD were some of the reasons for increased financial corruption. These shortcomings were also the cause of the continued dearth of efficiency and transparency in the policy-making process.

In order to make the whole debate comprehensible the book has been divided into two parts. In Part I, I have tried to discuss all those factors that have a vital contribution in the policy-making process. Chapter 1 comprises an analysis of the threat perception in the South Asian region during the period 1979–99. For this I have studied the Afghan crisis as it started in 1979 and its development until 1999. The crisis is discussed purely from a Pakistani perspective: what it meant for Islamabad; what the Pakistani leadership wanted to gain from it; and the manner in which it affected Pakistan's security. This is followed by a detailed analysis of the tension between India and Pakistan. Chapter 2 contains a detailed description and analysis of the official policy-making system and process. In this chapter I have tried to present both the system and the process purely as a mechanism employed by an organization such as a government to procure military hardware.

In Chapter 3 I have discussed at length various actors involved in the policy-making process, their relationship, interests, motives and the influence they exert on decisions. The actors have been categorized into two: (a) those who have a direct link and interest in weapons acquisition, and (b) others who play an indirect role. The second type of actor was found to be deliberately strengthening the position of the military in controlling the policy-making process.

Chapter 4 presents an analysis of the cost of military buildup for Pakistan. Considering the country's economic constraints the amount that Islamabad can and chooses to spend has a direct impact not only on the war preparedness of its armed forces but also on what the government procures. From 1982–90 Islamabad's burden was shared by Washington, who had decided to extend financial support to its ally to help it strengthen its military machinery in order to counter the Communist threat. Despite the external support, defense-spending from Pakistan's own resources tended to increase and after the arms embargo in 1990 the burden was entirely shifted to the country's economy. Successive governments have opted to maintain a high military expenditure because of the threat felt from India. In fact, this has been the pattern for the past fifty years of the country's history. In doing so, the policy-makers have shown no concern for the growth and development of the society and economy. The social underdevelopment, in fact, adds to the cost of military buildup. The strong pro-military lobby in the policy-making elite has not only discouraged any reduction in security spending but also not allowed the government to consider reducing the high percentage of financial wastage in the defense sector.

Chapter 5 discusses the policies of Pakistan's arms suppliers: the US, China, the UK, and France. Since the end of the 1950s the Pakistani military has developed an inclination towards American equipment. The manner in which Islamabad aligned its certain policies to facilitate American arms transfers during the 1980s and reasons for which the arms supply was interrupted again are discussed in this chapter. Traditionally, in the absence of American military hardware, Pakistan has looked to China, France and Britain. Beijing has always been instrumental in enhancing the quantity of Pakistan's equipment. The debate on Pakistan–China arms transfer links in this chapter looks at the subtle changes in the security links between Islamabad and Beijing.

Chapter 6 comprises an analysis of Pakistan's defense industrial complex. A description of the defense industrial infrastructure and an understanding of Pakistan's indigenous weapons production capabilities was considered necessary in comprehending some of the arms procurement decisions that were made.

Chapter 7, the first chapter of Part II, carries an in-depth analysis of numerous arms procurement decisions taken during the period from 1979–90. These decisions were related to both domestic and foreign acquisitions. Chapter 8 follows on from this and analyses decisions carried out from 1990–99. In Chapter 9 the nuclear proliferation decision-making has been studied. Although the country's nuclear program pertains to non-conventional defense, which is not the main subject of the book, the debate was included in the study for two reasons. First, the behavior of the policy-making process in this area was found to help understanding arms procurement decision-making further. Second, the development of the non-conventional defense capability had a direct bearing on the conventional arms procurement decision-making. Particularly after 1990, when the American government imposed an arms embargo on Islamabad and when Pakistan found it difficult to fulfill its military's major weapons modernization needs, policy-makers did not panic because of their peculiar confidence in nuclear deterrence.

The concluding chapter looks at the options that Pakistan has in terms of what it wants to and what it can possibly procure. The discussion in this chapter also contains an analysis of the probable changes in the domestic decision-making environment and the political leadership's ability to consider options other then hostile relations with India that may transform military planning at a strategic level. Any change in this level would impact on plans and related decisions at the operational and tactical level as well. What all this might mean for arms procurement policy-making is addressed in this chapter.

Part I

Like any policy-making process, the arms procurement decision-making of a country functions as a combination of multiple factors determining the final outcome. The number of actors involved in the policy-making process, their interests and influence, and the manner in which they exert this influence has a direct impact on decisions. Pakistan's case is no different. The official decision-making system, the domestic political and policy-making environment, industrial and technological capability, relations with arms suppliers, and political agendas that influence arms acquisitions policies are some of the factors that must be considered at the time of decision-making – not to mention the economic conditions that determine what a state can procure from domestic or foreign sources. Part I, comprising six chapters, aims at analysing all the elements having an impact on arms procurement and military buildup decisions in Pakistan.

1
Development of Threat Perception

Pakistan's adversarial relations with neighboring India play a vital role in the formation of its official threat perception and national security plans. Since 1947, the policy-maker's greatest concern has been to find means to thwart India's hegemonic designs or plans to gain a prominent status in the region's geo-politics. This was the basic framework in which the Soviet invasion of Afghanistan in 1979 was interpreted. The projection of the Soviet threat was vital for Pakistan's military modernization plans. This, however, led to embedding Afghanistan in Pakistan's security architecture. The real threat to Pakistan's security, which was more internal in nature, was a direct manifestation of Islamabad's Afghan policy. The fact remains that after 1979 Pakistan's primary source of threat continued to be India. The heightened India–Pakistan hostility culminated in increased tension and a nuclear arms race.

The Afghan crisis and Pakistan's security

The Soviet invasion of Afghanistan added a new dimension to the Pakistan government's threat perception. Officials presented the strategic development on the country's northern borders as a direct threat to Pakistan's security. Islamabad was of the view that, after consolidating its control over Afghanistan, Moscow would proceed further with the objective of reaching the 'warm waters' through Pakistan.[1] Although opinion on this perception was divided among the Pakistani leadership at the time, such a projection was considered expedient in order to cement US relations, which in the past had been impaired because of divergent views of Washington and Islamabad on Pakistan nuclear proliferation. In one respect, Pakistan's Afghan policy written in the early 1980s was a continuation of its earlier policy that viewed Afghanistan as a state inimical to Islamabad's security interests. The threat that it posed to Pakistan's security, however, was not comparable to that of India. With the Soviet troops taking control of Afghanistan, the government in Kabul was seen as a potent threat. The impact of Pakistan's Afghan policy on Islamabad–Washington relations was noticeable.

The only possibility of acquiring American military and economic aid was through convergence of views between the two countries. This chance was provided in the 1980s when Washington desperately searched for a formidable ally in South Asia to counter what was perceived as the Soviet threat to the free world. Like in the 1950s and 1960s, when American military hardware was transferred to Pakistan in bulk, the Reagan administration agreed to strengthen its South Asian ally through arms transfers and economic aid. The equipment transferred thus was a major contributory factor in formulating Islamabad's offensive posture towards the former USSR. Considering Islamabad's hostile relations with New Delhi, this equipment gave Pakistan the ability to stand up to its traditional adversary, India, as had happened in 1965. The war led to an American arms embargo imposed on both the South Asian countries, but Pakistan was the most affected because its dependence upon American equipment was greater than India's. The overriding factor was military modernization. Pakistan's then President, General Muhammad Zia-ul-Haq, was certain that an agreement with Washington regarding containment of the Soviet threat could help him obtain assistance from the United States. This, he or any other leader could not have hoped to get prior to 1979–80. It was a matter of how Islamabad played its cards. The Pakistani propaganda about Soviet plans dovetailed with the anti-Communist hype in the US that had grown especially after Ronald Reagan's ascendancy to power. In the US, the political mood favored the strategy to punish Moscow for what appeared to be the USSR's transgression of the norms of East–West relations laid down after the Second World War.[2] It was Cold War politics rather than a popular belief in America that its South Asian ally's integrity was at stake, and this led Washington to believe the Pakistani propaganda. According to Professor Stephen Cohen, who was working with the US State Department at the time, opinion in the American Congress was divided. Twenty per cent of members thought that the USSR's objective was merely the invasion of Afghanistan; 60 per cent thought it was to establish influence in the Gulf; and only 10–20 per cent were of the view that there were plans to invade

Pakistan.[3] The military government in Pakistan played a major role in playing up the Soviet military action as Moscow's ancient dream to control the warm waters. At that time Islamabad used every incident and twisted it to prove the USSR and its allies had launched insurgency operations to destabilize Pakistan and threaten its security. For example, activities of the terrorist organization *Al-Zulfiqar*, which had made Kabul its base, were deliberately interpreted and presented to the public and international community as an act of terrorism sponsored by Moscow.[4] The Soviet and Afghan Air Forces' violation of Pakistan's airspace was aimed at attacking *mujahideen* training camps and bases and was Moscow's reaction to Islamabad's involvement in low-intensity operations against its troops. This increased pressure on Pakistan but the government did not desist from the insurgency operations. Throughout this period, Pakistan enjoyed the blessings of Washington.

The USSR at no time intended to alter substantially the security architecture of the South Asian region. Its primary concern was to alter the wave of domestic political developments in Afghanistan that were detrimental to Moscow's strategic interests. The Soviet apprehension of political developments in Afghanistan had grown gradually. The government in Kabul had shown signs of its desire to move away from Moscow. Towards the end of the 1970s, the Shah of Iran had managed to woo Afghanistan's President Daud relatively away from Moscow. With the Islamic revolution in Iran in early 1979, the Soviet leadership had become more apprehensive regarding its impact on the Muslim population in the USSR's southern territories bordering Afghanistan.[5] The whole operation was certainly ill planned, and would not necessarily have culminated in the invasion of Pakistan.[6] Any threat to Pakistan came after General Zia had put his plans into action aiming at providing credibility to his peculiar claims. His Afghan policy led to an influx of migrants from Afghanistan, the majority of whom were, later, trained to carry out low-intensity operations against Soviet troops. There was a definite element of risk-taking in this strategy that led to Afghan and Soviet aircraft's airspace violations and bombardment of Pakistan's territory, which refreshed the memory of the famous U-2 affair during the 1960s. Even at that time, Islamabad was motivated by its desire to get quality equipment from the US which was forthcoming as compensation for its the support. Obtaining US assistance was the objective that had inspired the Pakistani military regime not to entertain the Indian invitation to Islamabad for joint diplomatic efforts to facilitate the withdrawal of Soviet troops from Afghanistan.[7] The mutual mistrust of Islamabad and New Delhi, and Pakistan's policy-makers' perception of the invasion as an Indo-Soviet conspiracy to eliminate Pakistan, made the acceptance of such proposal a distant possibility. Besides, Islamabad had always mistrusted Moscow because of its close links with India.

This is not to suggest that the crisis did not have any affect on Pakistan's security. The threat, however, was more a result of the policies pursued by Islamabad like the decision to allow free entry of Afghan refugees into Pakistan. Granted that the Pakistan–Afghanistan border is highly porous and it would have been difficult to curb all flow of population from the neighboring country, the access provided to the Afghan refugees by the Zia regime resulted in an unnaturally large influx of people. These refugees brought problems to Pakistan; their presence had a negative impact on the economy and ecology of the host state.[8] The menace of the narcotics trade and small arms proliferation anchored in society during the early 1980s was a repercussion of the free hand given to Afghan refugees by the Zia government – not to mention the corruption that increased within the military during this period.

Pakistan's Afghan policy, tailored and controlled primarily by the Pakistani military intelligence, sought to link the future of Afghanistan with Pakistan and *vice versa*. A pro-Pakistan regime in Kabul was a politico-strategic aim to be achieved at all costs. The relations forged with the various *mujahideen* groups during the military struggle in the 1980s put the GHQ at Rawalpindi in a favorable position to manipulate local politics in Afghanistan. The northern neighboring state was essentially viewed as territory that could provide the much-desired strategic depth to the Pakistani armed forces. Also, the trained Afghani manpower was seen as an additional infantry battalion to be used, if the need arose, against India. Furthermore, the control of Afghani politics was a central point of the broader plan to project Pakistan as a militarily strong Islamic country that would eventually control the newly established Central Asian republics and the states in the Middle East.[9] It was with this objective that Zia opposed the Geneva Accord that led to the Soviet withdrawal without the removal of General Najibullah.[10] His main objection was that the agreement did not allow the replacement of a pro-Soviet regime with one favorable to Pakistan's interests. The Army top brass pursued this objective even after Zia's death through ISI's covert operations in support of friendly but fundamentalist groups in Afghanistan. Help to such elements continued to shift, suiting Islamabad's convenience, from Gulbadeen Hikmatyar to the *Talibaan* who were trained by the Pakistani Army.

Through this extreme Islamic fundamentalist group, the Pakistan government hoped to enjoy total influence over the political situation in Afghanistan. This policy erected serious repercussions for Islamabad's relations with Iran. Tehran was conveniently excluded from the equation of determining Afghanistan's future. The *Talibaan* forces opposed the Iranian-supported Afghan commander, Ahmed Shah Masood. The Pakistani military intelligence agency, ISI, the Islamic fundamentalist party, Jamaat-i-Islami, and the *Talibaan* was the triangle fighting proxy wars and moved by anti–US sentiments. Islamabad's support to the *Talibaan*, who provided backing to the extremist, Osama Bin Laden, was a source of tension between Pakistan and the US. Islamabad's Afghan policy was also an example of the inability of successive governments to control defense and foreign policy-making.

India–Pakistan: a conflict of interest

For the Pakistani decision-making elite, India remained the primary concern. This attitude had its roots in the mistrust that resulted from the bitter historical experience of the partition of the subcontinent in 1947. A similar situation persisted on the other side of the border. What fuelled their bilateral hostility further was the manner in which the two states viewed each other: their military capabilities and their own standing in the international system. The vested interests ensured the bolstering of this tension for their political and, at times, personal and organizational gains. Tension grew markedly during the 1980s and the 1990s. The pattern of hostile bilateral relations developed during the 1980s led to heightened tension and culminated in an overt nuclear arms race between the two countries. One can only blame the policy-making elite of both countries for not finding peaceful solutions to the outstanding issues and taking recourse to military options.

India's strategic perspective

One of the many complexes of Indian policy-makers is that they find it extremely frustrating not to be treated as one of the top-ranking states.[11] The struggle to attain this objective is naturally reflected in the overall national and defense policies of the country. New Delhi's defense policy since independence has focused on India's recognition as a prominent regional, if not a global, power. This self-image was very neatly captured by Krishan Nayyar who believes that: 'The world has learned to live with US power, Soviet power, even Chinese power, and it will have to learn to live with Indian power.'[12] This ambition was thwarted by the presence of two 'villains': China and Pakistan. In their separate ways, both these states would not allow India to have the satisfaction of being recognized as a regional power.

Consequently India's defense posture underwent an evolutionary growth from the traditional power posture (1961–63), to latent nuclear deterrence or the extended power posture (1989–96).[13] Although the Indians assert that these changes were more focused on China, New Delhi was equally worried about Pakistan. In fact, major ground and air force deployments by New Delhi target Pakistan. Islamabad's constant effort to challenge New Delhi's authority over Kashmir and its military superiority, particularly with reference to the nuclear option, has always perturbed Indian policy-makers. From India's strategic standpoint, Pakistan is an irritant that must be taught a lesson or disciplined. The situation acquires a serious dimension when compounded with New Delhi's desire to assert itself as a regional power. This will keep India's smaller neighbors nervous.

Pakistan's strategic perspective

Pakistan's defense planning, on the other hand, revolves only around India which is seen as a powerful state with hegemonic ambitions. The popular feeling is that the Indian leadership has never been comfortable with the birth of Pakistan and will never relinquish its objective to reunite it with India. An alternative view is that its stronger adversary's urge is to turn Pakistan into some kind of a client state like the rest of the smaller countries of the region. While both these ideas are totally unacceptable to Pakistan's decision-makers and general public, it is also felt that the stronger neighbor has the military muscle to achieve such goals. The disconcerting factor is Islamabad's realization of the inequality between the two countries. This paranoia is conveyed to the general public with a similar situation on other side of the border. The resulting insecurity is further deepened by the fragmentation of Pakistani society. Successive governments have been unable to anchor a sense of nationalism in the country. They have instead resorted to some kind of consensus through focusing primarily on security issues such as nuclear deterrence and Kashmir, thus instilling the fear of India in the hearts of the general public. At times the growing ethnic and sectarian violence is blamed on a *foreign hand,* which in Islamabad's dictionary has invariably meant Indian or Israeli conspiracy to destabilize the only Muslim country in South Asia. Although foreign involvement in the present domestic crisis in Pakistan cannot be entirely ruled out, it is also a result of Islamabad's imbalance policies. Meanwhile, the military regimes and political governments operating under the influence of the armed forces have resorted to displaying a tough military posture against India for the purposes of gaining political legitimacy. Portraying Pakistan as the only regional country to stand up to India's 'bullying',[14] therefore, has always been the corner-stone of Islamabad's strategic perspective. Pakistan's defense program, in particular, is closely tied with India's. The maintenance of conventional balance and Islamabad's decision in 1998 to follow India in going overtly nuclear not only depicts Pakistan's sense of insecurity *vis-à-vis* New Delhi, but also its urge to hamper Indian efforts to gain a higher position in the hierarchy of nations.

Enhancement of threat in the Indian subcontinent

India–Pakistan rivalry can be categorized into two phases: 'passive' and 'active'. The first type indicates years marked with a certain lull in hostilities, while the second shows a reverse situation. Incidentally, one of the main features of the 'active' period has been the relative strengthening of Pakistan's military. Islamabad's ability to sufficiently modernize its equipment has normally resulted in its increased ability to challenge its adversary's security equation. The latest 'active' cycle of hostilities started from 1984 extending to 1999 and beyond. The exceptionally bad relations between India and Pakistan during this period were obvious from the series of threatening events. These developments depict the basic policy-making trend in the two countries pertaining to the solution of outstanding issues and other matters. It was obvious that their leadership believed in solving bilateral issues through military means. Situations were deliberately allowed to reach a crisis point; however, in all these cases, war was avoided. None the less, these conflicts provided the justification to both the countries to sustain their military buildup. Furthermore, these events left marks on the defense policies of both countries, with a greater impression on Pakistan.

1984: a year of crisis

1984 was the beginning of accelerated military activities in the Indian subcontinent. This was the year that India occupied the Siachin glacier followed by rumors of New Delhi's plan to attack its adversary's main nuclear facility at Kahuta.

The Siachin is one of the largest and longest non-polar glaciers stretching over approximately 1000 square miles and situated over 20 000 feet above sea level. It is located in the Karakoram mountains at the northern edge of the Indian subcontinent and south of China (see Map 1.1). The Siachin issue is an expansion of the existing Kashmir problem. After 1984 the glacier was converted into the world's highest battleground when India launched its operation Maghdoot to occupy the glacier. Over the years, the Indian expedition proved disastrous.

New Delhi's venture was essentially a pre-emptive measure to block Pakistan's control of the glacier. It is no secret that since the 1970s both countries had an 'eye on the glacier'. Their disagreement was due to their divergent comprehension of their bilateral agreement reached at Simla on the demarcation of the boundary and the Line-of-Control (LoC). Towards the end of the 1970s Pakistan had tacitly begun to exercise control over the glacier by charging a fee from mountain climbing expeditionary parties which annoyed New Delhi. In 1983 General Zia-ul-Haq ordered its occupation but the Army could not comply due to its involvement in the country's politics and Afghanistan. Hence, in 1984, Islamabad was practically caught with its 'pants down'.[15] It was felt that Pakistan would threaten Indian territory through its control of the glacier. Such a notion is debatable and the Indians admitted their strategic miscalculation and blunder at a later stage.[16] Equally questionable was the Pakistani propaganda after the invasion. The establishment believed that India had plans to threaten the Karakoram Highway linking China with Pakistan. Keeping in view the geographical and atmospheric hazards this argument also is far from logical.

One can only read the event in light of the outstanding Kashmir issue. The LoC drawn up as a result of the Simla agreement between India and Pakistan did not extend to the glacier; hence, the dispute arose over that who had legal authority over the territory. The Indians were apprehensive of a Pakistani plan to control the glacier that would be counted by New Delhi as its adversary's victory. The situation was reminiscent of 1965 when Pakistan had pre-empted India and won a partial victory in the Runn-of-Kutch. Military aid received from the US made Pakistan stronger than before, making India nervous about the development. The Indian military would have naturally wanted to offset any advantage to its rival; therefore, exerting its military strength was a logical path to follow. Turning Siachin into a 'bloody turf', which is of no military and strategic importance to either state, was a blunder that was later realized by the policy-makers on the both sides. Artillery exchanges and local skirmishes at regular intervals became a routine feature, costing the two countries dearly. As much as their leaderships might want to defuse the situation, their perceptions and

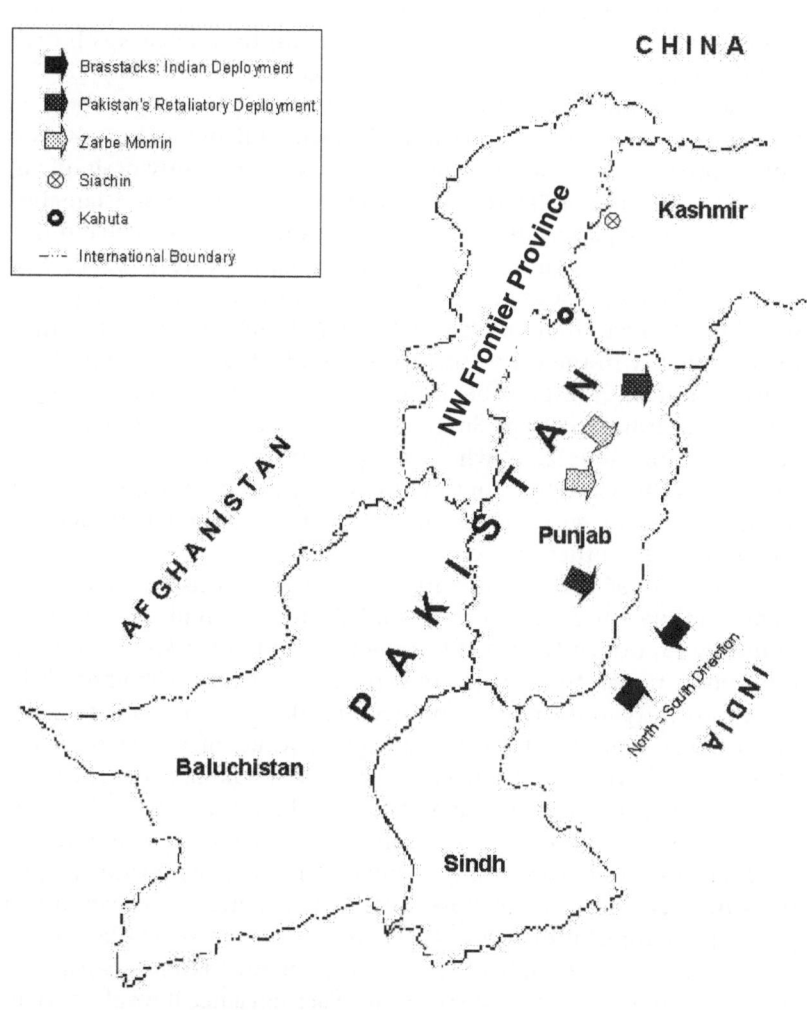

Map 1.1 Pakistan–India: Tension Points

mistrust of one other were major stumbling blocks in the way of resolution of the issue. Talks were formally initiated in 1986 but reached deadlock in 1992. Prime Ministers of India Indar Kumar Gujral and Nawaz Sharif of Pakistan, both sworn in 1997, showed an inclination to negotiate, but with little success. The main hindrance was India's desire to settle the issue to its own advantage.

It was the same year that Islamabad was caught by the rumors of an Indian plan to launch a 'conventional' attack on Kahuta. The news was not surprising. For New Delhi it would have been highly desirable to eliminate the core of Pakistan's nuclear program. By 1984 Islamabad had sufficiently propagated its non-conventional defense capabilities. Islamabad's nuclear program would essentially create a 'balance of terror' in the region and put it on a par with India. This development was indeed unwelcome in New Delhi. The Israeli attack on the Iraqi nuclear facility in the early 1980s was a model that India might have opted to follow. A responsible Indian source indirectly confirmed the rumor. According to Jasjit Singh such an action was originally suggested by the Americans who later passed on this information to Pakistan. In his view, by adopting this strategy, the US had thought of 'killing two birds with one stone'.[17] This was a far-fetched idea not endorsed by any other source.

The rumors of such an Indian plan generated a lot of tension in Islamabad, which gave rise to a similar threat in reply. General Zia promptly cautioned New Delhi of the possibility of Pakistan's retaliatory conventional strike against Indian nuclear facility at Trombey.[18] Its close proximity to the city of Bombay was likely to cause phenomenal havoc. This 'balance of terror' strategy is what, in the view of the Pakistani authorities, dissuaded New Delhi from taking extreme action. Kahuta, being close to the Indian border, could not have been defended effectively so the most feasible option was to force New Delhi to realize the high cost of such a venture. Islamabad, it must be remembered, by then had its F-16s, which possessed deep penetration capability to pop pop the threat into action. Pakistani sources described Islamabad's response as a fine example of crisis management.[19] From this perspective the crisis was a useful experience for Pakistan since it would have to employ a similar strategy in another two years.

Interestingly, the fear of an attack on Kahuta recurred in July 1998 prior to Pakistan's nuclear tests. This time the Pakistani authorities claimed that a number of Israeli aircraft were spotted in India. It was believed that New Delhi, in collusion with Tel Aviv, wanted to destroy Pakistan's nuclear facility, thus making it difficult, if not impossible, for the latter to conduct its nuclear tests. There was no confirmation of the presence of Israeli fighter-jets in India.

1986–87: Operation Brasstacks

India tried to exercise its military supremacy again in 1986–87 over its two most important neighbors, Pakistan and China, by holding its military exercises, Brasstacks and Chequerboard, close to their borders. Both exercises generated a fear of war in the countries concerned. The first one was held close to Sindh, the politically turbulent southern province of Pakistan (see Map 1.1).

New Delhi had multiple objectives in mounting Brasstacks. These ranged from the assimilation of weapons procured during the early 1980s and co-ordination and training of the military in handling the two main strike corps that formed the main offensive force against Pakistan, to improving the Indian population's faith in the defense of the country. A source speaking for New Delhi termed the exercise as an indigenous version of *glasnost*.[20] None the less, the main reason behind this seemingly innocent military operation was to scare Pakistan and possibly to dissuade it from covertly supporting the insurgency in the Indian Punjab. The fact that the Indian troops were deployed close to the Pakistani border with live fifth line ammunition was sufficient to scare Islamabad.

Brasstacks created panic in Pakistan's military circles. The fear was further deepened by New Delhi's attitude: the Indian Director-General (Military Operations), when quizzed by his Pakistani counterpart about the use of live ammunition, was suspiciously evasive.[21] This helped to confirm the views of certain members of General Zia's rather small defense decision-making circle that India would try to break Pakistan again as it did in 1971, or else she was using this as a ruse to divert Islamabad's attention while Indian troops invade Pakistani controlled Kashmir. Lt General (Retd.) Hameed Gull, who was one of the important corps commanders at the time and promoted later as the head of ISI, told the author that Lt Generals K.M. Arif and Akhtar AbduRehman supported such an idea.[22] General Gull expressed his dissatisfaction with their findings. According to him, Islamabad had totally misinterpreted Brasstacks. He further explained that had India meant to launch a war it would not have left the strategic points to its north vulnerable. Islamabad tested the hypothesis that India did not intend to wage a war through counter-deployments. A division was moved to the area in Punjab opposite the Fazilka-Akhnoor and Gurdaspur-Pathankot sectors. This move unnerved New Delhi and the exercise was called off.

This result did not change Islamabad's approach towards Brasstacks, which Gull blames on the incompetence of Arif and AbduRehman. He further asserted that the two generals had never commanded a corps; thus, they had no experience of planning or interpreting military operations. Being close aides to Zia, their opinion prevailed upon the opposition from others. The Indian view was that Pakistan's propaganda was basically to justify its arms procurement and nuclear program. The nuclear deterrence factor, which was a joint Zia-Arif strategy, was indeed important and was used during the exercise to offset the Pakistani military's conventional weapons inferiority during Brasstacks.[23]

The Indian military exercise exposed Pakistan's strategic weaknesses and India's military strength, especially the force it could muster. From then onwards, it was obvious that Islamabad would plan in terms of offsetting its adversary's conventional force advantage through non-conventional defense. Islamabad's publicity of its nuclear program, especially official statements, aimed at sending a message across the other side of the border. With a history of tense relations, any Indian statement was understandably insufficient to assuage Pakistani apprehensions emanating from Brasstacks. Fundamentally, Pakistan's response to the operation signified its policy-makers' determination to 'pay its adversary in the same coin'; an attitude that was to be re-endorsed by Islamabad in 1989.

1989: Zarbe Momin

Soon after Operation Brasstacks the Pakistani armed forces launched their own large-scale exercise, Zarbe Momin. From a technical and strategic standpoint, it sought to try out the new strategy of 'offensive-defense' and assimilate the $5 billion worth of weapons that had been procured during the first half of the 1980s from China and the US.[24] The main idea, nevertheless, was to send a message regarding Islamabad's capability and intention of defending its territory in case of Indian aggression. The exercise was held in the south of Punjab, between the Chenab and Indus rivers, around the districts of Multan and Dera Ghazi Khan, approximately 250 km from the Indian border (see Map 1.1).

The exercise area was divided between the 'Blue-land' and 'Fox-land'. The first was designated as Pakistani territory and the second as the enemy area. It was presupposed that there was domestic political turmoil in 'Fox-land' that was to be used by the 'Blue-land' forces for launching an offensive. The entire strategy, documented later in a restricted paper called the 'Gulf Crisis 1990', was based on capitalizing India's political vulnerability on its eastern border of the Punjab province. It was hoped that the Sikhs would join hands with the Pakistani military in fighting the Indian armed forces.[25] Such an understanding was based on the political turmoil in the province. The Zia regime had indeed provided support to Sikh dissidents. Unable to counter India through conventional forces, the only feasible option was to weaken the adversary domestically. This would naturally dampen the enemy's capability to launch an offensive. New Delhi has always blamed Pakistan for the turbulent situation in the Indian Punjab. This plan was primarily the brainchild of the then Pakistani Army chief, General Mirza Aslam Baig. The strategy was used again in the 1990s when Islamabad encouraged insurgency operations by the Kashmiri freedom fighters in the Kashmir valley. The system of layered defense was introduced in the exercise focusing on the use of less manpower and increased firepower.[26]

Zarbe Momin failed to impress either the enemy or the analysts at home. Despite the fact that the 'Blue-land' forces took the counter-offensive in 'the area of their choosing', logistics and communication did not match up even in the initial phases.[27] There was also a lack of co-ordination between the infantry and mechanized forces which was one of the reasons the exercise had to be terminated abruptly. Moreover, the Indians were not too convinced of what the Pakistani military had learned in terms of firepower and mobility.[28] According to a Pakistani Army general, it was a 'degenerated military exercise' that mainly presented General Baig's publicity venture.[29] This notion, though not entirely incorrect is an over-simplification. The benefit reaped from the exercise was non-military in nature in that the Pakistani defense establishment had asserted itself politically by taking a major decision to take the war into the 'enemy territory', which could provide the government with more 'room' to negotiate.[30] The military dimension of the plan could not be ignored either. One of the lessons learnt from previous wars was that Pakistan did not have the capability to defeat India. Islamabad must, however, remain in a position in which it could negotiate peace with the adversary at less favorable terms. Hence, it was vital to gain control of some part of the enemy territory. Furthermore, it presented a Pakistan that now had the weapons to launch an offensive into enemy territory. This was a definite departure from the strategy of the past.

Kashmir crises: the gathering storm

In 1990, the bilateral relations between the two neighbors grew tenser over the Kashmir issue. This coincided with the publication of an article by the American journalist, Seymour Hersh, claiming that both

India and Pakistan were on the verge of a nuclear war that year. He further asserted that Pakistani F-16s carrying nuclear devices were on standby. Although American diplomats serving in the region refuted his argument,[31] the situation in the subcontinent was far from normal. There was a definite increase in 'saber-rattling'. The international community was alarmed because this surge in tension coincided with an increase in nuclear activities in the region. A nuclear arms race with 'Pakistan crossing the line in 1990' was not a welcome idea.[32] India, in particular, was uncomfortable with the linking of nuclear deterrence with the Kashmir issue. In pursuing the nuclear option, New Delhi did not want to take along the baggage of any outstanding issue since it would make the program controversial and strategically objectionable to the international community. Islamabad's policy was just the reverse: Pakistan's aim was to get the world to realize that peace in the region was threatened unless the issue was resolved. After the two wars with India in 1965 and 1971 there was a certain consciousness that Kashmir would have to be forced on to the international agenda and this would not be possible unless a sense of urgency could be generated.

The Kashmir imbroglio, which dates back to 1947, picked up speed in 1989 when the people of the Kashmir valley began to protest against the oppression of the state (provincial) government supported by the central government in New Delhi.[33] The Indian government firmly believed that Islamabad was directly involved in fanning the insurgency in order to facilitate the breaking away of Kashmir.[34] This was an argument that the Indians continue to maintain. This notion was based on the fact that, since 1947, the Pakistani ruling elite had made Kashmir the *raison d'être* of Pakistan, and linked it with the survival of Islam in the South Asian region. The accusation was not entirely incorrect except that the crisis was not of Pakistan's making.[35] The origin of the crisis was purely internal. In fact, the developments in the Kashmir valley that started in 1989 were so sudden that they had taken the Pakistani Army by surprise. The movement was initiated by the Jammu and Kashmir Liberation Front (JKLF), which in any case did not support the idea of joining Pakistan. The party's pro-independence stance was equally unacceptable to Islamabad. According to a senior Pakistan military officer, the Army saw it as an opportunity to keep the Kashmir issue on the front 'burner' but had initially mishandled the situation owing to lack of clarity in the position they should take regarding the JKLF. It was later that they began to support the Islamic fundamentalist but pro-Pakistan groups such as the *Hizb-ul-Mujahideen*.[36]

Since the re-emergence of the Kashmir issue in 1990, relations between India and Pakistan grew further hostile. After 1990, Pakistan's main offensive was diplomatic in nature. This accompanied Islamabad's support to Islamic fundamentalists from all over the world who came to the region to aid the Kashmir insurgency movement. In the early 1990s the Army had yet not planned an offensive the way it had done in 1965. Allowing non-state actors to incite turbulence in Kashmir, as it was hoped, would force India to consider holding a plebiscite in the valley giving its people the right to choose between joining Pakistan or India. Dr Mehboob-ul-Haq, a Kashmiri by origin and who had a standing in Pakistan's top decision-making circle, told the author in an interview that in a plebiscite, which would be jointly held in both Pakistani and Indian Kashmir, the majority was likely to decide in Pakistan's favour.[37] This may prove to be an extremely generous assumption on his part, even though it reflects thinking in the Pakistani camp. The two neighbors have also worked at times on the bidding of the international community to negotiate on the issue and talks were resumed in 1997 by their newly installed governments. With the change in the political scene in both countries, one cannot expect a solution without a softening of the rigid attitude held by the two. Unfortunately, the option of giving the Kashmiri people a third alternative, that is, to gain independence from both India and Pakistan, was never considered seriously.

Unlike in 1965, Islamabad did not start the insurgency in Kashmir in 1989 but the Army generals saw the continuation of the domestic political disturbance as beneficial to Pakistan's military objectives. The question, however, is how long would the Kashmiri insurgents be able to fight the Indian armed forces? Although disturbed by the continuation of tension and hostility in the valley, towards the mid-1990s, the Indian leadership had started to grow more confident of its position in the disputed territory. How much risk Pakistan can afford to take in reversing the situation, or how long can India prolong the non-settlement of the issue are the questions worth asking.

The military's own calculations of resolving the issue were laid down in July–August 1999 when a covert operation was launched to unsettle Indian troops controlling strategic peaks at Kargil. This operation, which was launched with the help of the Kashmiri and Afghan *mujahideen*, was reminiscent of 'Operation Gibraltar' of 1965 when the Pakistan Army again had tried to use a military option to resolve the issue. The idea was to capture certain strategic peaks left vacant by the Indian troops during the winter season and the Pakistan Army would gain control of the area and put the adversary in a disadvantageous position. It was hoped that, eventually, India would be forced to take a decision to negotiate a settlement of the Kashmir crisis which would be favorable to Pakistan. As mentioned earlier, it was assumed that once India agreed to hold a plebiscite in Kashmir the results would be favorable for Islamabad. Unlike 'Operation Gibraltar', which had led to a medium-intensity conflict with the adversary, the Pakistan Army felt confident that the nuclear capability would deter India from reacting aggressively the way it had done in 1965. Caught between a low-intensity conflict situation and the possibility of a nuclear war, India would be forced to solve the Kashmir issue.

The plan drew a blank, New Delhi resorting to a more hostile stance against Islamabad. In the intense bilateral skirmishes that ensued, an unarmed Pakistani Naval surveillance and ASW aircraft was shot down by the Indian Air Force in August 1999, killing sixteen officers on board. The Indian authorities claimed that the plane had violated their airspace and was shot down in Indian territory while carrying out spying activities. Surveillance activities carried out by Pakistani Naval aircraft, which are French-built Breguet Atlantic aircraft fitted primarily for ASW operations, were nothing out of the ordinary and are missions normally carried out by the navies. However, the timing of the flight was peculiar: the PN could have avoided sending its aircraft so close to the LoC at a time when tension at the Kargil front had not yet subsided.

The Pakistani authorities claimed otherwise given that the wreckage was found in Pakistan's side of the LoC. This and other hostile actions added to the existing tension, with the international community worried about a possible eruption of a full-scale war between the two countries. Appeals were made to both India and Pakistan to desist from aggression but to no avail. Prior to this, the Pakistani air defense had shot down two Indian aircraft violating Pakistan's airspace in Kashmir with one of the two pilots captured alive handed over to the Indian authorities. The naval aircraft was downed to avenge the shooting down of the Indian aircraft but was also designed to make good publicity for the Indian Prime Minister, Vajpai, and boost the morale of Indian troops. The IAF was given a fair beating by the Pakistan Army's Air Defence Command. Contrary to the wishes of New Delhi, which thought it could punish the insurgents through intense air raids, the IAF incurred losses and humiliation. The internal psychological dynamics of such adverse actions was indeed a crucial factor, but there was also another psychological pressure that was likely to have caused the shooting down of the PN aircraft. This was that the Indian intelligence agencies and the military had been severely admonished for not noticing the infiltration of the *mujahideen* at Kargil. Hence the urgency to thwart any territorial or airspace violation.

The military maneuver at Kargil failed owing to international pressure exerted on Pakistan to withdraw support from the insurgents. The Pakistan Army blamed the political government for the subsequent withdrawal. After taking control of the reins of government, however, the same military managers that had planned the operation in the first place announced a unilateral withdrawal of troops from the forward deployment. There was, none the less, no change in Islamabad's stance in providing diplomatic support to the Kashmiri people engaged in a struggle against New Delhi. It is not likely that the Pakistan Army would repeat this venture in the near future unless it musters the technological capability to take on India's superior military might.

The conventional arms race

The military competition between the rivals, as expressed by their conventional arms race and overall military buildup, grew intense during the 1980s after American arms transfers to Pakistan. It is not that an action–reaction phenomenon in the Indian subcontinent was generated after these transfers but the arms race factor became more prominent after 1982. The fresh transfer of military technology to Pakistan was highly displeasing to the Indians who did not want its adversary to adopt any degree of offensive posture. A Pakistan battered and bruised after the 1971-war suited Indian strategy, whereas a more aggressive Pakistan, in combination with Beijing, disturbed these plans. New Delhi's irritation over the US aid to Pakistan was expressed by Mrs Gandhi in these words: 'I am told we cannot possibly object to Americans giving Pakistan what it is in need of, but we also feel [that] they are being armed to an extent which is well beyond their needs'.[38] Islamabad was accused of fuelling the arms race in the region and trying to disturb the regional military balance established primarily by India in 1971.[39] Table 1.1 shows the equipment disparity between India and Pakistan before and after 1979 and is indicative of the military competition prevalent in the Indian subcontinent.

By the mid-1980s, India had superiority over Pakistan in terms of manpower (ratio 2:1), divisions (2:1), main battle tanks (2:1), surface ships (4:1) and combat aircraft (3:1).[40] In light of the above facts and figures, 'India's superiority in conventional arms is unquestioned.'[41]

Table 1.1   Military Technology: a Comparative Analysis

	India		Pakistan	
	1979	1988	1979	1988
Total Manpower	1 104 000	1 362 000	450 600	480 600
Combat Aircraft	614	714	220	338
Tanks	2 120	3 250	1 300	1 600
Artillery	1 350	2 295	1 353	416
Submarines	8	14	6	6
Frigates	24	24	0	0

Source: The Military Balance, 1979–80, 1988–89.

Some analysts, however, have a problem in using only the quantitative approach. Raju Thomas, for example, feels that numbers do not say much about the qualitative balance.[42] This notion voiced India's apprehension of certain military hardware acquired by Islamabad after 1979, the most significant being the 40 F-16s procured from the US. Although the procurement of this multipurpose, agile and deep-penetration fighter aircraft did not make Pakistan any stronger, New Delhi was obviously disturbed by the acquisition. It reacted by procuring more aircraft aimed at offsetting PAF's technological advantage. Even if one were to discard the simple bean-counting methodology for an assessment of comparative technological capabilities, New Delhi maintained a definite edge over its adversary. Figures given in Table 1.2 help in ascertaining the technological advantage enjoyed by IAF.

Prior to 1982, the Pakistan Air Force was dependent on Chinese and old American technologies that were no match for the Indian Air Force's aircraft inventory. The qualitative dimension was denoted by a limited number of different versions of French Mirage III (33) and Mirage V (62). During the 1970s, Islamabad also increased the number of aircraft through acquiring about 200 F-6s series from China although the combat capability of these aircraft was limited. India, on the other hand, had 300 MiG-21s, 10 MiG-23BNs, 13 MiG-23UMs, 50 Su-7BM/KUs, and 16 Jaguar GR-1s in her inventory with 85 Jaguars, 62 MiG-23BNs, 13 MiG-23UMs, and eight MiG-25s on order.[43] It was this technological disparity that the PAF's top brass had hoped to decrease by acquiring technologies such as the F-16s and AWACS. This equipment would have acted as a major force multiplier tilting the military balance in Pakistan's favor to a great degree. Negotiations were conducted with Washington for the transfer of AWACS on the grounds that they were needed to beef up security against Soviet forces. The US government looked at the prospects seriously but decided against it because of the cost factor and the implication it would have on the South Asian regional military balance. The technology of these aircraft based on an advanced real-time command, control, communication, and Intelligence (C3I) system could be used by the PAF in keeping track of India's forward deployments. Such advance information of the enemy's military moves could definitely have improved Pakistan military's overall defense in qualitative terms. Other aircraft obtained by the PAF during this period were either refurbished or of Chinese origin, and in any case were inferior to those India possessed. In 1996–97 the IAF procured the Russian Su-30 fighter aircraft that were sufficient to counter any technological advantage enjoyed by the PAF in the past.

Islamabad could not manage to carry out major refurbishment of weapons from American assistance for its other two services, the Army and Navy. Other transfers from the first aid package worth mentioning comprised the American TOW anti-tank missiles, shoulder-fired Stinger missiles, and Cobra attack helicopters which the Pakistani Army needed to enhance its capability to retard the speed of the Indian armoured corps onslaught. The service could not manage to improve the quality of its tank force, which was primarily of Chinese origin, and some American tanks procured from the first aid package were secondhand/refurbished M-48As. Furthermore, the program for the indigenous manufacture of tanks did not take off, leaving the Army with a weak 'defensive' posture. The quality of Pakistani tanks, moreover, was not comparable with India's. The progress made by India in the field of ballistic missile technology denoted by the different types of missiles such as *Nag, Prithvi, Agni* and *Trishul* added to the concern of Pakistani generals. Islamabad reciprocated by developing its own missiles although the Chinese M-11 missiles procured by Pakistan, however, constituted the most potent weapon.

The Pakistan Navy did no better during the 1980s. Owing to the low priority afforded the service, it could not benefit from the first aid package. Three anti-submarine warfare (ASW) capable aircraft, P-3C Orions, were procured for the service from the second package but these aircraft were withheld as a result of the arms embargo. The aircraft were finally released after the Brown Amendment in 1996, but the Navy was not satisfied with their capability and the condition in which they were handed over. As part of the arms embargo, the manufacturer was not allowed to release certain critical spares nor provide training to enable

Table 1.2 India and Pakistan: the Qualitative Difference in Combat Aircraft Technology

Aircraft type	Country	Quantity	No. of engines	Per engine thrust	Total thrust	Combat weight	Thrust/ weight ratio	Max ordnance load	Range (combat radius)
F-16 A	Pakistan	34	1	23 830 lb	23 830 lb	22 785 lb	1.046	15 200 lb	604 nm
Mirage III	Pakistan	63	1	13 668 lb	13 668 lb	21 164 lb	0.646	8 818 lb	647 nm
Mirage V	Pakistan	56	1	13 668 lb	13 668 lb	21 825 lb	0.626	8 818 lb	675 nm
Mirage 2000	India	35	1	21 384 lb	21 384 lb	23 545 lb	0.908	13 889 lb	800 nm
Sea Harrier	India	20	1	21 500 lb	21 500 lb	14 052 lb	1.530	8 000 lb	400 nm
MiG 19	India	N/A	2	7 275 lb	14 550 lb	16 755 lb	0.868	1 102 lb	370 nm
MiG 21 FLs ' Fishbed'	India	74	1	13 613 lb	13 613 lb	17 240 lb	0.790	1 102 lb	704 nm
MiG-21 MF	India	80							
MiG 21bis	India	170	1	15 653 lb	15 653 lb	19 235 lb	0.814	4 409 lb	795 nm
MiG 23BN	India	54	1	24 728 lb	24 728 lb	36 926 lb	0.670	6 613 lb	728 nm
MiG-23 MF	India	26							
MiG 25 R	India	8	2	24 691 lb	49 382 lb	81 570 lb	0.605	6 614 lb	1 006 nm
MiG 27 M	India	148	1	25 353 lb	25 353 lb	39 903 lb	0.635	8 818 lb	291 nm
MiG 29 A	India	74	2	18 298 lb	36 596 lb	33 598 lb	1.089	6 614 lb	810 nm
Jaguar SEPECAT GR. Mk 1/T.Mk 2	India	109	2	8 040 lb	16 080 lb	24 149 lb	0.666	10 000 lb	460 nm
Su-30	India	40	2	27 557 lb	55 114 lb	51 257 lb	1.075	13 228 lb	1 620 nm
Shenyang F-6	Pakistan	100	2	7 165 lb	14 330 lb	22 046 lb	0.650	1 102 lb	370 nm

Table 1.2 India and Pakistan: the Qualitative Difference in Combat Aircraft Technology (continued)

Aircraft type	Country	Quantity	No. of engines	Per engine thrust	Total thrust	Combat weight	Thrust/ weight ratio	Max ordnance load	Range (combat radius)
FT-6	Pakistan		2	7 167 lb	14 334 lb	22 046 lb	0.650	1 102 lb	370 nm
FT-5	Pakistan	30	1	5 952 lb	5 952 lb	11 905 lb	0.500	1 102 lb	378 nm
F-7	Pakistan	79	1	13 448 lb	13 448 lb	16 603 lb	0.810	2 205 lb	324 nm
A-5 Fantan	Pakistan	49	2	8 267 lb	16 534 lb	20 913 lb	0.791	4 409 lb	324 nm

Sources: The Military Balance, 1996–97, Encyclopedia of World Military Aircraft, Vol. I & II.

the Navy to carry out operational flying of the P-3Cs. Eight American frigates were obtained on lease but these were called back in 1989–90 leaving the Navy with a gap to be filled by the country's own resources. The service purchased three different types of major weapon systems from 1990–96 but these hardly enhanced the PN's capabilities to a degree that would constitute a serious threat to the Indian Navy's 'blue-water' capability. The Pakistan Navy continued to maintain a defense posture. It hoped to ensure security of its coastline through cheaper equipment such as the Chinese gun-boats and missile-boats. The desire to enhance the service's traditional defensive posture to that of safeguarding the EEZ seem never to be fulfilled. The French embargo of 1999, which temporarily stopped the transfer of an Agosta 90-B submarine, exposed PN vulnerability in even maintaining a credible defensive posture. It must be noted that most of the major weapons systems in the PN inventory were of western origin; thus, spares support could be cut-off at any time.

Purely from a military technology standpoint, Islamabad's acquisitions after 1979 did not reverse or substantially narrow the military capabilities gap with its adversary. Given Pakistan's serious resource constraints this was difficult to achieve. The US military assistance provided it with an opportunity to modernize some of its major equipment, which in relative terms made the country more 'muscular' and put it in a position from which it could challenge the Indian calculation of a military balance in the region. It was not that F-16s made Pakistan stronger than its adversary but this acquisition made it more robust than India would want. The foreign assistance, unfortunately, was not a permanent feature of Pakistan's military modernization program: the end of the Cold War, followed by the break-up of the Soviet Union, caused the drying up of the American source of arms supply to Islamabad. In fact, one of the constant worries of the Pakistani leadership has been the *ad hoc* nature of arms transfer relations with the US, as opposed to India, which enjoys a far more stable linkage with its main arms supplier, Moscow. New Delhi's links with its primary supplier was a result of their successful diplomatic relations. Islamabad, on the other hand, could never gain significance for its arms suppliers to have permanent arms transfer links with Pakistan. Islamabad was mainly to be blamed for not streamlining its foreign, economic and other policies to establish more stable ties with the US or any other state. It was in these circumstances that the nuclear option was formally introduced. The decision-makers' predicament was to find a solution for countering a stronger adversary in a situation where national economic constraints did not allow military modernization.

2
The Official Decision-Making System

A formal decision-making system is established to achieve certain goals specified by the policy-making authority. The primary goal of these structures is to help management make rational decisions based on predetermined objectives. The system then carries out these designated activities and its performance is ascertained on the basis of how well it has obtained the goals. Any deviation is considered as a malfunction. This description applies to policy-making systems of all sorts of organizations, government and corporate, with a caveat that decision-making is ridden with personal biases, politics, and other factors. The main focus of this chapter is to look at Pakistan's official arms procurement decision-making system, its structure, and the manner in which the system is required to function.

Background

The predecessor of the present decision-making system dates back to 1947 when a defense decision-making process was established after the creation of the country. This system had its roots in the pre-independence colonial decision-making, the process of centralized military authority in colonial India was brought about in 1902 when Lord Kitchener was made Commander-in-Chief and the sole military advisor.[1] What made his power even more significant was that the seat of the main decision-making authority, the British government, was not in India itself. This system, with powers vested in the commander-in-chiefs of the three services, continued even after the country's independence. In theory, the control of policy planning and execution lay with the then newly formed Cabinet Committee for Defense (DCC) (see Figure 2.1).

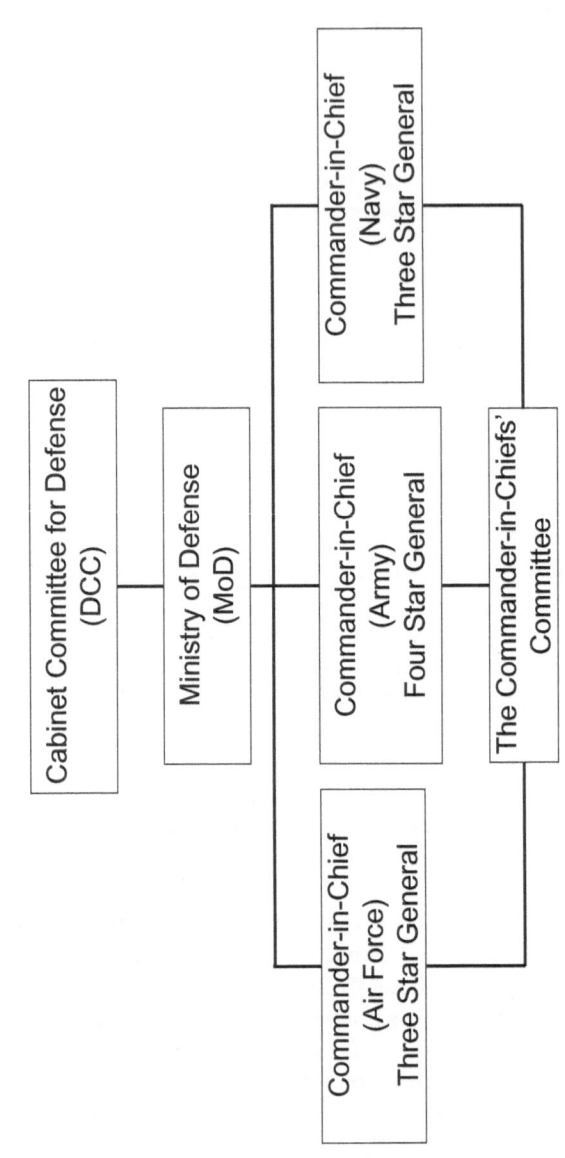

36

Defense Command and Control

Cabinet Committee for Defense (DCC)

Ministry of Defense (MoD)

Commander-in-Chief (Air Force) Three Star General

Commander-in-Chief (Army) Four Star General

Commander-in-Chief (Navy) Three Star General

The Commander-in-Chiefs' Committee

Figure 2.1   Defense Command & Control

Its membership was only extended to political representatives who were members of the parliament; the service chiefs, who were under the administrative control of the Ministry of Defense, could only attend the meetings. In reality its power was limited, which was due to structural and other flaws in the system. Policy matters pertaining to national security planning did not fall within its ambit. This problem was compounded by the civilian decision-makers' lack of knowledge of military affairs, hence encouraging the service chiefs to exercise their independence. This practice reflected the peculiar thinking common to many Third World countries where the political or diplomatic policies of a regime are divorced from military planning.

The structural problems were also prevalent at the level of the armed forces where the system suffered from a lack of joint planning. The only joint forum for the three services was the Commander-in-Chiefs' Committee (CCC) headed by an officer of the rank of Brigadier serving as the secretary. The rest of the staff was posted from the services and were accountable to their parent organizations while they were serving on the committee. The function of the CCC was confined to simple administrative tasks. Each Commander-in-Chief was responsible for the planning and control of his own service but the Army's larger size, and its active involvement in the country's politics, put it in a dominant position over the other two services, creating an imbalance in the military system and a lack of co-ordination among the armed forces. Compounded by the relatively weaker position of the political governments, which failed to act as a central control authority, there were the fiascos of 1965 and 1971. In the first instance, the Army launched a full-scale military operation called 'Operation Gibraltar' without consulting or involving the other two services.[2] This operation involved dropping troops behind 'enemy lines' in the Indian occupied Kashmir with the objective of helping a civil war type situation. The plan not only failed to achieve its objective, but also resulted in a military encounter between India and Pakistan. In the second case, the Navy came to know of the beginning of the war in 1971 through an Indian radio broadcast.[3]

The 1973 White Paper on 'Higher Defense Organization'

The political government of Prime Minister Zulfiqar Ali Bhutto, installed in 1971, tried to remove the aforementioned discrepancies. A new system was introduced through a policy paper in 1973. There were two clear objectives of the government: (a) change the civil–military balance in favour of the former, and (b) inject the concept of joint planning for defense in the policy-making process within the military. One of the primary concerns identified in the White Paper regarding the old system was that it had become totalitarian in nature as it was set out to serve the interests of a single service, the Army.[4] It tried to eradicate this through major structural changes in particular those in the command structure of the armed forces.

Provisions were made to anchor this structure in the constitution of the country. Article 243 (1) of the 1973 constitution laid the responsibility for national defense with the head of the government and the chairperson of the Cabinet Committee for Defense, who, in both the cases, was the Prime Minister. The cabinet on the whole was responsible to the parliament.[5] The Commander-in-Chiefs of the three services re-designated as Chief-of-Staffs now came under the control of the Prime Minister. This system was further amended in 1988 placing the President as Supreme Commander (see Figure 2.2). The Prime Minister was to be assisted in his task of defense decision-making by a Minister and/or a Minister of State for Defense. Other changes were also made such as shifting the naval and air headquarters to Rawalpindi and Islamabad – where the seat of power was located. This system continued until the 1980s when modifications were made through the eighth amendment to the 1973 constitution in 1985. After this, the President acquired the authority in defense decision-making;[6] the rest of the system, nevertheless, was not changed.

Management of defense decision-making

One of the major features of the new system was its hierarchical structure spread on three tiers. It also created a multi-divisional and complex bureaucratic structure for the input, implementation and feedback related to dedicated defense decisions.

Higher management level

This new 'command and control' concept, which resulted from various studies carried out under the supervision of Bhutto's regime, put the civil government at the top of the ladder (see Figure 2.2). All decisions were to be taken and endorsed by the Cabinet Committee for Defense (DCC) which represented the nucleus/core of the decision-making apparatus. There are thirty-two committees which report to the cabinet committee. The DCC deals with policy matters at a strategic level and its meetings are chaired by the constitutional

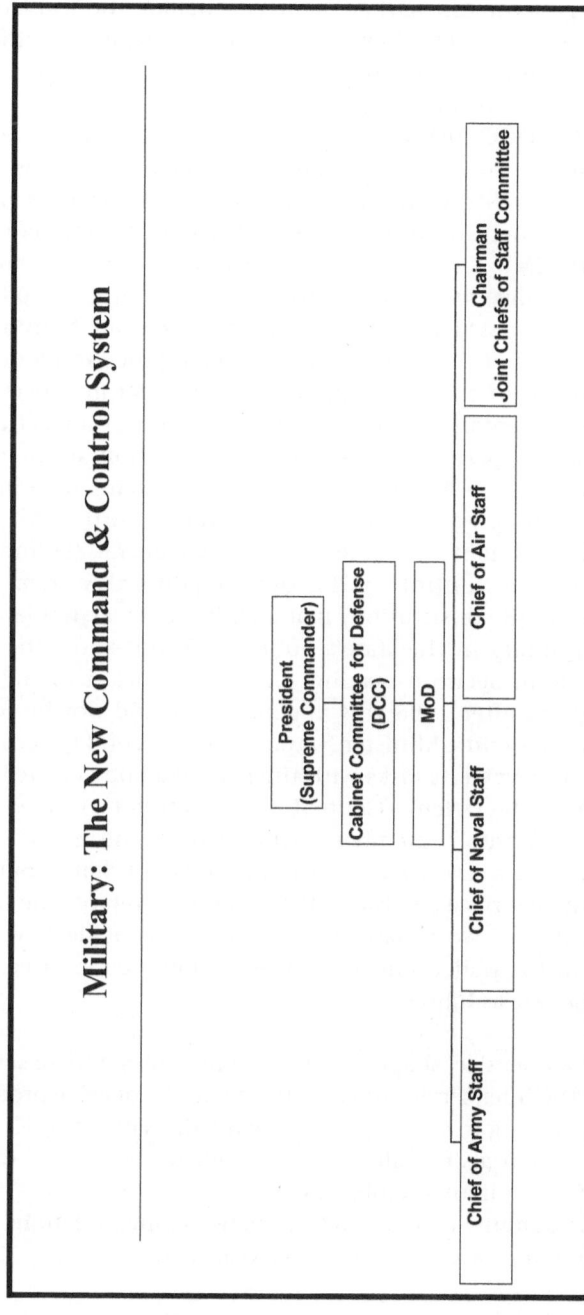

Figure 2.2   Military: the New Command & Control System

head of the government. Other members include the Ministers for Defense, Foreign Affairs, Commerce, Frontier Regions, Industries, Finance, Communications, and Interior, with their secretaries, the Chairman JCSC. Although no military personnel could become members of the DCC they were, however, allowed to attend or could be summoned to the meetings. These membership rules, as can be seen from later events, could not stop a military official becoming a head of the Cabinet Committee. This was when General Zia became the head of the DCC. The imposition of Martial Law by General Zia in 1977, therefore, gave rise to an anomalous situation in the top level of the policy-making structure. Under this situation, an Army general became a member of the DCC by virtue of being the President of the country, and was in charge of appointing the other members of the committee in accordance with the powers vested in him as the head of the state and Supreme Commander. This situation, as mentioned earlier, was dealt with by making the President the Supreme Commander through amending the constitution. Article 58 (2B) of the eighth amendment was aimed at helping General Zia find a way to consolidate his control in the overall political system of the country. The 1973 constitution, made with an understandable bias against the military in the state's politics, did not allow the armed forces any role in decision-making. Despite the desire of successive political regimes after 1985 this amendment could not be revoked until 1997 during Prime Minister Nawaz Sharif's second government.

The DCC performed its tasks with the help of a body created in the 1970s known as the Defense Council, which would be headed by the Minister for Defense. Its members included the Ministers of Finance and Foreign Affairs, the Chairman Joint Chiefs of Staff Committee (JCSC), the three service chiefs and the Defense Secretary. The Defense Council's nature of work has tactical bearings, as it deals with the monitoring of the policies devised by the cabinet committee. Other responsibilities are as follows:

(a) review the role, size, shape and development of the three services;
(b) monitor the inter-service organizations and the defense production facilities keeping in view the recommendations of the JCSC;
(c) consider proposals for allocation of funds to the various organizations of the military establishment;
(d) annually examine, review and formulate recommendations for the approval of the DCC, or present five years defense plan; and

(e) formulate policies regarding the induction of new weapons systems.

The achievement of the two parliamentary bodies, however, is debatable. Since their creation, there have been no public reports pertaining to any controversial debate in the parliament regarding any decisions that have been taken by successive governments. Furthermore, the three service chiefs have become a permanent feature of the DCC with all input provided by the armed forces.

Middle management level

In the newly designed decision-making hierarchy the Ministry of Defense represented the first level of the middle management, it being the main organization responsible for the implementation of the policies formed by the top decision-making management. It also provides feedback to the DCC on the implementation of decisions. Moreover, it serves as the top management's channel of communication and control of the armed forces. This philosophy seems to have been borrowed from Britain where the Ministry of Defence and the office of the Chief of Defence Staff were created to deal with inter-service rivalry and improving government's communication with the military.[7] Considering the nature of Pakistani power politics, however, there has not been a single instance where the MoD has played any part in resolving issues emerging from inter-service rivalry. The Ministry of Defense is a formal bureaucratic agency responsible for the handling of all matters pertaining to defense. It is comprised of a fairly large workforce of civil servants, which includes some military personnel as well. Its organizational structure is 'multi-divisional' with the main organization divided into smaller independent units looking after different areas of activity.

This structure had its inherent problems, such as the creation of 'pluralism', which means that the presence of a large number of administrative and functional units increase the overall management cost, and the inefficiency of the system was ensured through poor communication between the various units. The Ministry's bureaucratic structure was overly expanded owing to the creation of two independent core divisions: defense and defense production, the latter division being created under the aegis of the 1973 White Paper. It was hoped that such expansion could help in facilitating the performance of a wide range of activities. The two divisions were further sub-divided but only the defense production division will be discussed here because of its direct relevance to the subject.

However, before proceeding further I wish to describe another bureaucratic establishment that facilitates arms procurement – the Ministry of Finance's special wing, dealing with the provision of resources for weapons acquisitions and other military plans. This wing also carries out the monitoring of funds allocated for the purpose. This branch, known as the Military Finance wing formulated on the same hierarchical pattern as other departments (see Figure 2.3) analyses the financial aspects of the requirements of the Ministry of Defense which are related to its various activities and branches. It is also responsible for providing information to the government through the Minister of Finance (who is a member of the DCC) on the financial dimension of arms procurement activity. Last, but not least, it has responsibility for authorizing every expenditure.

A supporting organization is that of the Military Accountant General (MAG) which acts as the payment and monitoring authority for the military. One of MAG's officers, the Controller Military Accounts (Defense Purchase), is responsible for the payment of all acquisitions made by the defense establishment from internal or external sources. The rest of the organization carries out the tasks of preparing and auditing the accounts of the military establishment. One of the lacunas of this department relates to its administrative control. Although its officers are on the payroll of the Auditor-General of Pakistan, they draw their salary and other allowances from the defense budget and are dependent upon the defense establishment. This has proved to be a major hindrance in carrying out a fair check of these accounts. It must be noted that Pakistan's Audit Department (PAD) does not conduct a legislative audit, and its budget is not independently approved by the parliament. This not only reflects on the poor quality of work done by the MAG's establishment in keeping a check on military expenditure, it also makes the authenticity of its system of financial checks and balances questionable.

Defense production division

The name given to this particular section of the Ministry does not provide a clear idea of the wide range of activities it carries out. It deals with weapons acquisitions from both foreign sources and within the country (see Figure 2.4).

The Defense Production Division was structured using the multi-divisional and project oriented approach. The sub-divisions related

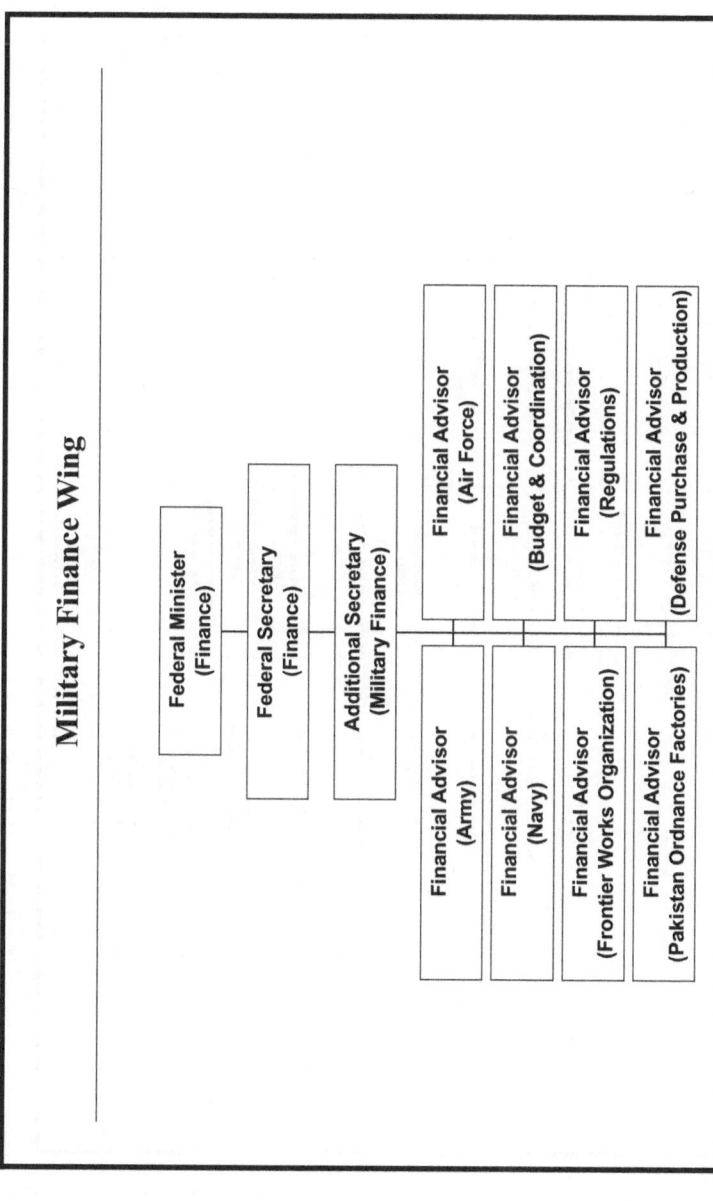

Figure 2.3   Military Finance Wing

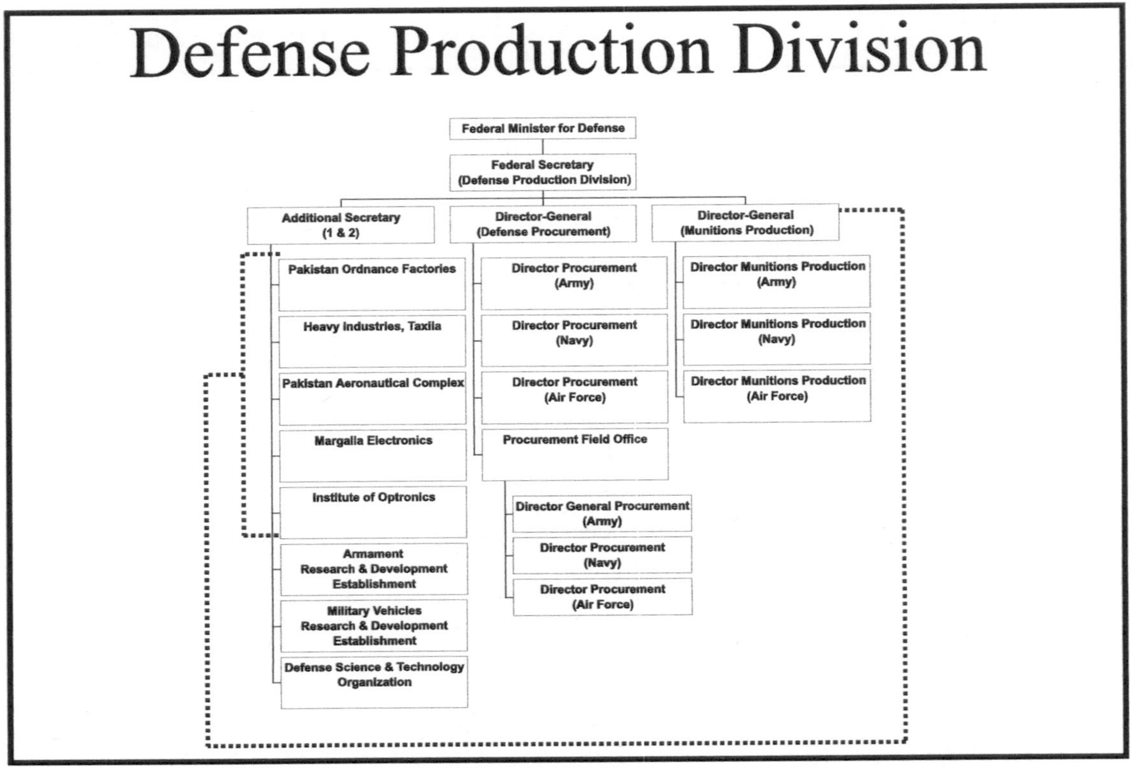

44

Defense Production Division

Federal Minister for Defense

Federal Secretary
(Defense Production Division)

Additional Secretary
(1 & 2)

- Pakistan Ordnance Factories
- Heavy Industries, Taxila
- Pakistan Aeronautical Complex
- Margalla Electronics
- Institute of Optronics
- Armament Research & Development Establishment
- Military Vehicles Research & Development Establishment
- Defense Science & Technology Organization

Director-General
(Defense Procurement)

- Director Procurement (Army)
- Director Procurement (Navy)
- Director Procurement (Air Force)
- Procurement Field Office
- Director General Procurement (Army)
- Director Procurement (Navy)
- Director Procurement (Air Force)

Director-General
(Munitions Production)

- Director Munitions Production (Army)
- Director Munitions Production (Navy)
- Director Munitions Production (Air Force)

Figure 2.4   Defense Production Division

to smaller organization or offices located at the Ministry of Defense headquarters and its field establishments. The field establishments again are of two types. The first category consists of the two sub-organizations of the director-generals defense procurement and munitions production. The second type consists of the POFs, Wah, Heavy Industries, Taxila and Pakistan Aeronautical Complex, Kamra, which are directly under the administrative control of the additional secretaries and are bigger facilities. This type of administrative arrangement, nevertheless, is rather difficult to comprehend. Government officials are of the view that such an administrative structure was made keeping in mind the principle of checks and balances, but this objective could be achieved through a less complicated system.

The sub-organization of the Director-General Defense Procurement deals with the work related to all types of procurement of weapons. The Director-General's activities include floating of tenders for procurement of military hardware, registering arms contractors and representatives of arms manufacturing companies, carrying out negotiations with the supplier companies, analysing the technical and cost specifications of the bids that are received in that office, and finalizing the deals. The three directors working at the MoD are involved with major weapons procurements. The field establishment, on the other hand, deals with spares, components and smaller routine procurements. The Army's dominance over this system was obvious. It is interesting to note that the officer representing the Army in the field organizational set-up is of the rank of a major general as opposed to his two counterparts who are of the rank of commodore and air commodore. This, it was explained, was because the office of the Director-General Procurement (Army) was handling procurement of greater financial magnitude than the other two. The other sister organization, Director-General Munitions Production, focuses on defense production activities and also exercises administrative control over five facilities highlighted in Figure 2.4. There are three facilities that were created to carry out research and development activities related to indigenous defense production. Not all of them are headed by officers from the armed forces. Organizations such as DESTO and IOP have scientists, who are civilians, as their chief executives.

Lower management level

This level consists entirely of the military. It comprises two basic parts: the JCSC and the three services. The Joint Chiefs of Staff Committee (JCSC) was created as a result of the 1973 White Paper. The objectives behind creating this organization were as follows:

a) to establish an integrated planning system;
b) to devise a system that was aimed at making optimum use of the limited resources available for arms procurement and overall national security;
c) to adopt a balanced approach to weapons procurement which would not focus entirely on strengthening one service only, and
d) to increase the civil government's participation in the defense decision-making process.

The bureaucratic structure given in Figure 2.5 was created in 1976 and, for the first time, provided the country with a thoroughly integrated inter-service mechanism for the higher direction of war. The Chairman of the Committee is a four-star general and his job description includes planning for war during peacetime and providing joint direction during war. In time of war, he is also to assist the Prime Minister as his principal staff officer on the general direction of the war. The Chairman can command no troops and the only military personnel under him are his staff. In addition, he carries out co-ordination with all inter-services organizations such as the Inter-Services Intelligence and the Pakistan Ordnance Factories. He is not authorized to interfere with the day-to-day running and direction of the armed forces. This particular injunction has limited the influence of the Committee. It also means that the Chairman cannot assert himself with the Army while it is directly ruling the country. The reasons for inserting a negative clause will be discussed in the next chapter.

All strategic and arms procurement planning is supposed to be carried out by the JCSC. It is responsible for processing weapons requirements in the light of the overall strategic plans that are made at the joint staff headquarters. One other consequential feature of the committee's role in defense planning is that it acts as a forum (for the three services) where a 'consensus decision' is arrived at. This looks similar to the Chinese decision-making style of 'fragmented authoritarianism' geared towards consensus decisions.[8] In Pakistan's case, the focus on consensus grew out of the top policy-makers' desire to stop fragmentation and to avoid wastage of resources. The stress on the 'optimum use of resources' mentioned in the 1973 White Paper required a system where the wastage of funds owing to the duplication of weapons procured, and so on, could be reduced if not eliminated entirely. This is an objective pursued by the decision-making authorities of other countries as well. For example, the American concept of 'management-by-objective-technique' introduced by the Packard

47

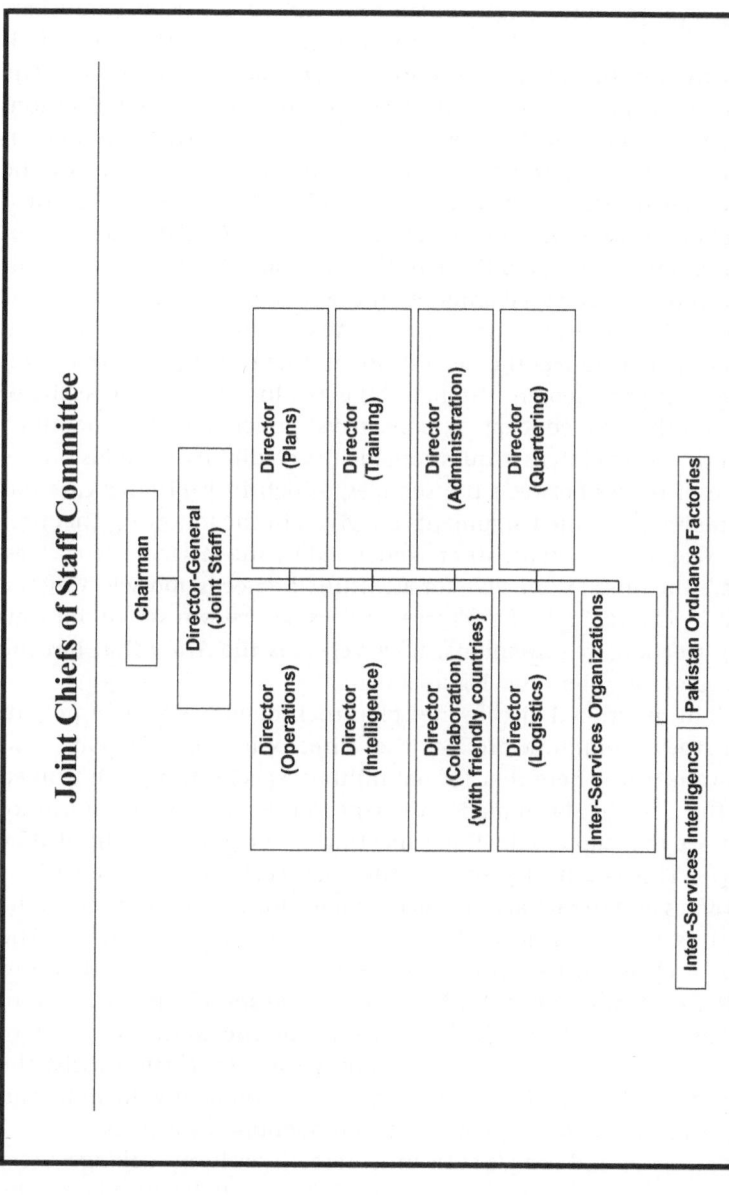

Figure 2.5 Joint Chiefs of Staff Committee

Commission, set up by President Reagan in 1981, was aimed at reducing the wastage during the arms procurement process.[9] It is for this purpose that the three service chiefs are required to discuss and debate the arms requirements of their respective services and the entire military in the meetings held at the joint staff headquarters (JSHQ). The agreed upon 'priority' list must result from a unanimous decision. Efforts were made during General Zia's rule to further strengthen JCSC's influence in the policy-making process. This refers to the Presidential directive issued on 20 March, 1985, which laid down instructions as to strict adherence to the principle that all procurement planning was to be routed through the Joint Staff Headquarters.[10] This order can be viewed as a comment on the functioning of the system as well. The directions given in the 1973 White Paper not being adhered to made Islamabad feel the need to streamline the importance of processing arms procurement through the JCSC. In any case, the reality of how smoothly the concept of 'consensus' policy-making functioned within the armed forces is questionable. Its credibility depends on the balance of power between the services, which in Pakistan's case has been missing. (Detailed arguments are given in the following chapter.)

The weapons requirements are generated by the services themselves. They are the ones responsible for assessing the needs and selecting the type of hardware. Each of the three services has its own system for generating weapons requirements; however, it is the Army that has the most elaborate system (see Figure 2.6).

Its eight functional directorates plan their demands for equipment, which they communicate to the directorate of Weapons and Evaluation.[11] It is here that formal military specifications are worked out. After a list has been made, it is sent to the 'Priority' committee of the service that supposedly holds meeting on a regular basis headed by the CGS. This committee screens the lists received from the various directorates and then makes its own final list for communication to the JSHQ. It is also at this level that a weapon system is 'prioritized.' The term is used when the Army high command expresses its desire to acquire a particular system with certain special specifications and from a specific source. The sub-organizations of the Master-General of Ordnance and Inspector-General Training and Evaluation assist the various directorates including the combat development wing in formulating arms procurement planning and induction of weapons.

The Air Force and Navy's arms procurement decision-making system is less complex than that of the Army. This is basically due to the smaller size and comparatively less complicated command structure

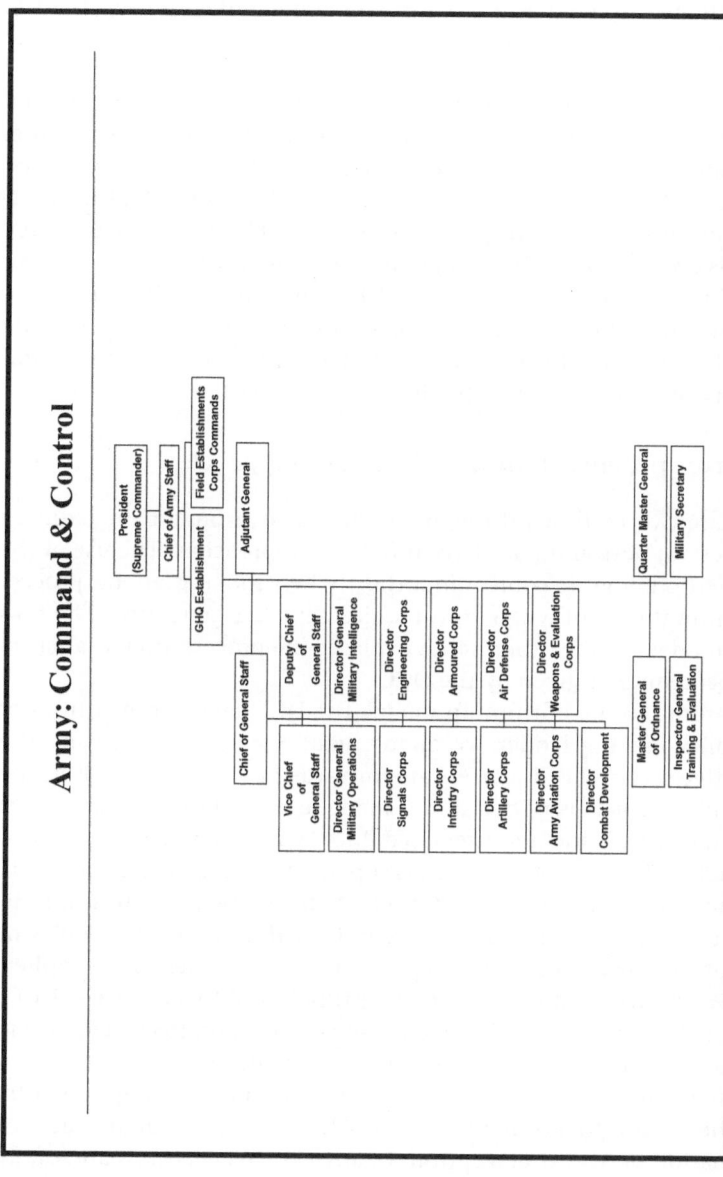

Figure 2.6 Army: Command & Control

(see Figure 2.7). In 1999 the Navy carried out extensive restructuring. This comprised the creation of a separate office of the DCNS (Plans). The ultimate objective was to leave procurement decisions to DG DP. This change, however, could not be made and the service continues to follow the old procedures.

Arms requirements for the two services are the responsibility of their plans and operations branches,[12] which, headed by the Deputy Chief of Air and Naval staff respectively, make lists of weapons required by their services. These lists are then finalized and approved by the 'priority' committees of the respective services to be forwarded to JSHQ. There is a variation in the procurement system of the smaller services at the implementation level. The PAF has an Inspector-General for the purpose while the Navy has no formal organization to carry out the task of the final checking of equipment prior to induction and the responsibility for this rests with fleet headquarters.

The arms procurement decision-making process

There are two methods through which the government's machinery operates: (a) bottom-up and, (b) top-down approaches. The use of the first methodology becomes important when looking at the process right from the point when arms requirements are generated. The top-down model, on the other hand, explains the process after a sanction for procurement is given by the DCC.

According to the bottom-up model, the DCC falls at the bottom of the policy-making ladder. Weapons requirements generated by the respective service headquarters are forwarded to the JSHQ. It is at the JCSC that a 'consensus' decision is made regarding the military needs of all the services and an 'Integrated Priority List' is prepared. This list, essentially a final list of weapons, is supposed to be made after a variety of issues have been considered, such as, threat perception, the military's capabilities, the military strength of the adversary, reports of trials of the weapons systems, report on the willingness of a supplier, and availability of funds. The final destination of this list is the DCC. The Cabinet Committee and its thirty-two sub-committees then debate the list or the overall defense needs in light of the information obtained from its members representing the different departments/ branches of the government. The DCC has no independent source of information on threat perception. In any case, its decisions are totally dependent on data provided by the MoD and the various military intelligence agencies, primarily the ISI. A policy decision by the

51

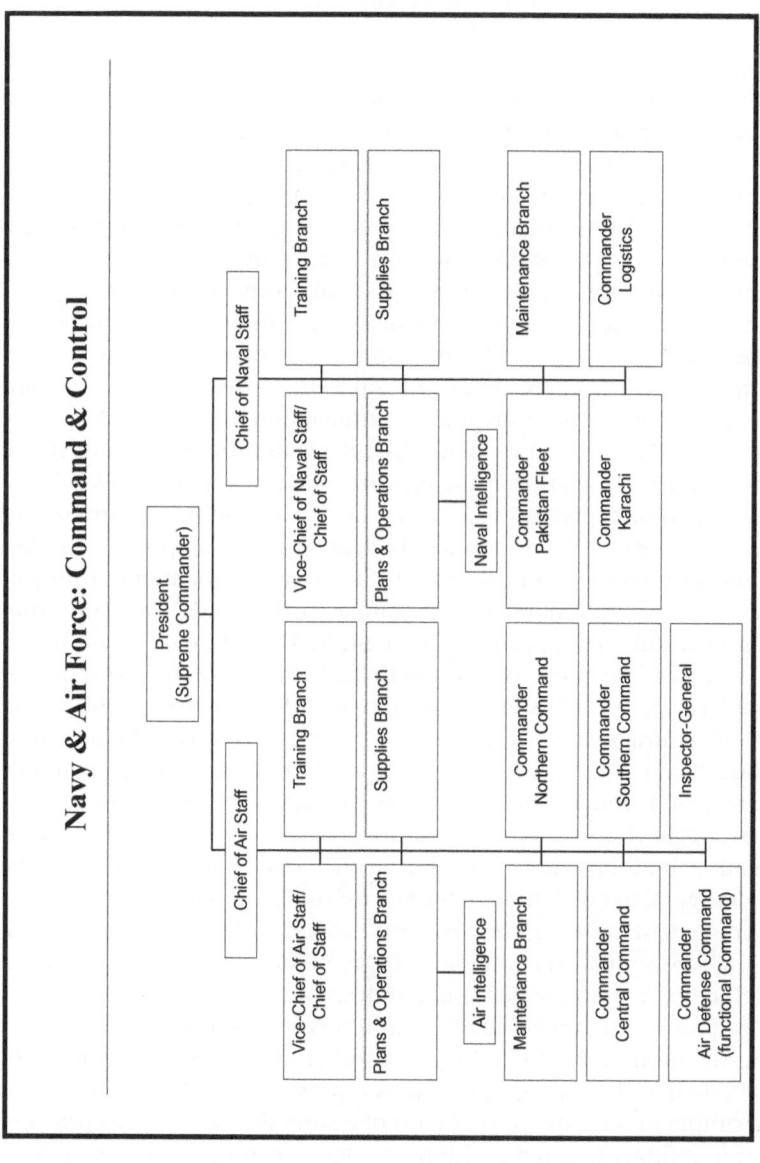

Figure 2.7 Navy and Air Force: Command & Control

government triggers off a chain of actions, which is linked with the implementation of the policy. This stage of the process can be better explained by using the top-down approach.

The cabinet committee's decision is communicated with the necessary instructions to the ministries concerned especially the Ministry of Defense. At this level, the necessary details, such as floating of tenders, conducting of negotiations, settling on the price, and so on, are carried out. The deals are signed at the Ministry of Defense from where the information is then passed on to the lower management level. The military receives the information through the joint staff headquarters and so begins the process of the induction of the procured weapons.

The government's approval can take two forms. In the case of short-term, ranging from 1–5 years, or urgent requirements, the decision-makers at the top can allocate funds immediately. In the second instance, long-term plans ranging from 5–10/15 years or where certain capital-intensive projects are under consideration, the approval can take a form of a statement of the government's commitment towards them. It must be kept in view that the procurement budget allegedly consists of three components: disbursements, carry-over liabilities (carried over to the next year) and commitment budget. The last component signifies the government's commitment to a certain procurement/project without actually producing the resources at the time. In certain other cases, governmental approval can be limited to a verbal commitment. For example, despite a commitment to the construction of another naval base in the 1950s, it has been argued that the funds took a long time in coming and construction work for this naval base, situated west of Karachi at Ormara, only started in the 1990s. This particular behavior is explained through Chaffee's model. According to this 'linear sequential strategic decisions model', strategic decisions are different from operational/administrative decisions. Policy-makers at the top make the long-range plans which are later left to middle or lower management for implementation.[13] Hence, the final implementation can take a long time. This is similar to Lindblom's theory of 'disjointed incrementalism' which reflects on implementation of decisions.[14]

The process described so far is not very complicated, and it deals with the method in which the government, with its multiple agencies and branches, determines the various inputs. None the less, it ceases to be a simple process in another kind of a situation where arms procurement or military buildup decisions are dependent upon assistance from foreign sources. The complexity in this case emerges from two basic factors. First, military modernization, as a result of foreign aid, cannot be seen as a constant factor in the procurement planning of any country. This should rather be viewed as representing an 'exception to the rule'. The decision-making system, therefore, has to be re-formatted to cater for the new situation. Second, if the weapons procurement funding is in the form of loans/grants that have not been specified as free, or the procurement is against some sort of trade barter agreement, then the recipient state has to work out the financial details concerning the repayment of both the principal amount and the interest, prepare amortization schedules, and incorporate all these details into the national financial planning framework. This generates an additional burden for the bureaucratic machinery. In such cases, the foreign policy objectives of the country have to be taken into consideration as well. An assessment has to be carried out on the opportunity cost of acquiring weapons through such means. For example, the Pakistan government's decision in 1981–82 to accept the American aid package at the commercial rates of interest was, it was argued, to take the aid without compromising the country's diplomatic independence.[15] There is also the issue of what the supplier in such a case is willing to transfer. Islamabad had to conduct negotiations with Washington for reaching an agreement on the weapons list. According to Lt General (Retd.) K.M. Arif, it took time and lot of hard work for the Pakistani team to get the US administration to agree to certain arms transfers.[16]

The arms procurement funding process

Provision of funds for the purchase of weapons is an essential part of the policy-making process. It is the Military Finance wing of the MoF that is responsible for carrying out the cost analysis, and keeping track of the funds. The Ministry on the whole is placed in the middle where it is approached independently by the DCC and the MoD. The allocation of grant assigned for defense during a financial year is further allocated by the MoD through the Military Finance wing located in the MoD building.

The provision of funds for procurement can take two forms. In the first instance, payment for a particular purchase is made at the request of the chief executive of the state or head of government as part of a special grant. The main reason for this is that the annual defense budget is packed with other expenses pertaining to personnel and maintenance costs that do not leave enough funds for fresh procurements. Moreover, some of the projects are too sensitive; therefore, any related transactions are not made out of the allocated budget. The nuclear program falls into this expenditure category. The second case relates to the annual defense expenditure, which is worked out according to a prescribed formula, and the services are allocated their share accordingly. The published annual budget is expressed in Pakistani rupees and whatever is spent out of the foreign exchange is converted in local currency before being booked as part of military expenditure.[17] The one flaw with this system is that because the allocations for major weapon systems and certain other projects is not made out of the annual defense budget, it leaves a lot of room for off-budget financing. For thorough discussion on this subject see Chapter 4.

At a glance Pakistan's official arms procurement system looks foolproof. It was designed to take care of ordinary and extraordinary situations. Its efficiency, none the less, depends on the policy-making environment and the general political culture of the country. Furthermore, there is the question of the number of actors involved in the process, their interests and the manner in which they exercise influence at the time of decision-making. There is also the element of the bureaucratic environment and behavior, and the information that is passed on to the higher management. Although even in a democratic political environment there are opportunities for manipulation of information by middle and lower management, these become increasingly possible in a less democratic system like Pakistan's. In addition, where the command and control of military bureaucracy by a civil government is not effective, it can result in creating more loopholes in the policy-making process. Whether this system has managed to deliver the desired objectives will be evaluated in the next chapter.

3
Pakistan's Power Politics and Defense Decision-Making

Decisions reflect the fundamental motivations and interests of actors involved in the policy-making process. The influence they exercise during decision-making is derived from their position and authority in the overall power politics of the state. It is a combination of all these elements that constitute the policy-making environment. In Pakistan's case one finds involvement of various actors that have been catagorized: direct and indirect.

Direct actors

This category comprises actors who have a direct interest in military buildup. These players include military and civil bureaucrats, arms suppliers and top decision-makers in the government hierarchy. Their actions have an impact on the policy-making process and the decisions taken. The relative significance of these actors varies according to the influence they enjoy in manipulating and controlling not only the relevant decision-making but also the overall policy-making apparatus of the state.

The military establishment

The reality of arms procurement decision-making is different from the prescribed system in which the top policy-makers of the government are supposed to exercise maximum authority in such matters. In Pakistan it is actually the military establishment, particularly the senior echelons of the armed forces, that set the defense policy and arms procurement agendas.[1] This situation differs from that of some developed countries, which have been studied so far. The military is not one of the many organizations of government struggling for the fulfillment of its goals, but the most powerful institution that was termed as 'the backbone of the nation'.[2] The military is the country's most powerful institution. It is the largest organized force with approximately 700 000 personnel. Compared with this, there seems to be no single democratic institution in Pakistan that can claim to have this number of members. In any case, the state's democratic system and institutions are weak.

The primary reason for the military's emergence as the most influential element in defense decision-making lies in its significance in the country's power politics. It assumed the responsibility of guarding the Islamic ideological identity and frontiers of the country. The Pakistani policy-makers have always had a fear of India; this anxiety has a serious ideological orientation. The idea is that India with its predominantly Hindu population cannot bear the existence of an Islamic Pakistan; and it is only the defense establishment that can provide the security of this ideological state. Projection of threat from India is fundamental to the survival of the Pakistani establishment that even views internal insecurity as a continuation of the external threat. Islamabad has also looked at its domestic political turmoil as the doing of a 'foreign hand.' India was never explicitly named but official circles tend to find the Indian intelligence agency Research and Analysis Wing (RAW) responsible for the sectarian and ethnic violence. This exercise of mutual accusations, in fact, became one of the features of bilateral relations between the two neighbors. The government's projection of threat does not take into account the domestic dimension of the political unrest. It does not say anything about the inability of the leadership to handle the domestic crisis and the internal forces which increase such problems. One cannot entirely rule out India's hand in the unrest in Pakistan; still, one can neither ignore the fact that such publicity is done mainly with the objective of convincing people of the need for military preparedness. Throughout the 1980s and 1990s India's projection of its military capability has undoubtedly undermined the confidence of neighboring states. This image, at times, was used in order to provide justification for a similar military buildup by Islamabad. Moreover, there is the outstanding Kashmir issue between the two, the existence of which has blocked any possibility of peace and confidence-building between Islamabad and New Delhi. The Pakistani leadership has time and again publicized the significance of Kashmir for the existence and survival of the country, hence its pledge to achieve the objective of eventually getting the territory 'back' from the Indians. An effort by the Nawaz Sharif government to improve relations with its adversary after 1997 was clouded by the military pushing the Kashmir issue on top of the negotiation agenda. Furthermore, resolution of this crisis depends upon a major change in the stance of both New Delhi and Islamabad, which at present does not seem possible. Its continuation, on the other hand, will tend to nurture the military competition between the armed forces in the Indian subcontinent.

It was with this mind-set that Islamabad embarked upon increased allusion to the threat from India and the need to respond to the perceived intimidation by the neighbor. The national security imperative makes it vital for the armed forces to act as a protagonist in political and national security affairs, as well as acquire weapons – a need which is acknowledged by the decision-making elite.

Military personnel, especially the top brass, are highly motivated to play their part in defense decision-making. This role is vital for procurement of armament. Apart from the strategic angle there are organizational, political and personal dimensions for the military's interest in weapons modernization: hardware provides justification for infrastructure expansion; arms imply creation of facilities for training, repair and maintenance, which require manpower; more hardware means an increase in human resources, which in turn increases the strength and influence of the defense organization. Equipment is vital for maintaining the political power of the armed forces which was acquired primarily as a result of weak democratic institutions and the inefficiency of politicians. The vacuum created by the inefficiency of political leadership was always filled by the military.[3] The top brass of the armed forces believes that it is the politicians (the civilians) who are responsible for the political crisis in the country and the military is a savior. This view was upheld even by the Supreme Court in the case of *Nusrat Bhutto vs. Chief of Army Staff*. In this writ petition, submitted in 1977, the Supreme Court had invoked Kelsen's law of necessity to justify Martial Law.[4] Unfortunately, this resentment of civilians, at times justified, has seeped into the junior cadre of military officers as well.

From a political angle the relationship between the acquisition of weapons and the power image of Pakistan's armed forces is typical of military-dominated Third World countries. In Pakistan's case, one cannot miss the political symbolism of military modernization, particularly when it is carried out from foreign sources such as the United States. Weapons transfers or any military cooperation with Washington is held as American support for the ruling party in Islamabad. Considering the state of the socio-political fragmentation of the country, both military and democratically elected regimes have looked outside for support. Within the largest constituency, the armed forces, American support has represented a popular symbol mainly because of Islamabad's dependence upon a third party to help manage its tense relations with India. Therefore, military buildup carried out particularly with the help of the US not only signifies American support for a particular government but has also become the measure of a regime's stability. In this respect, Pakistan's military governments were comparatively more fortunate: major weapons refurbishment occurred twice as a result of American aid which was received during two martial law regimes. These weapons, which signified the Western superpower's support for the otherwise shaky military regimes, provided them with the strength to survive – a strength they would lack internally. The contacts developed between the military establishments of the two countries during these arms transfers were beneficial for the US in helping Washington develop some influence over the thinking and inclination of Pakistani defense forces. This advantage was lost after the arms embargo was imposed in 1990.

The political and organizational imperative, at times, was found to be more instrumental than the national military strategic needs in determining military planning. The Army's involvement with the civil government's operation of weeding out 'ghost schools'[5] or temporary induction of 35 000 Army personnel in the Water and Power Development Authority (WAPDA) were indicators of the organizational building instinct of the armed forces. The inability to develop a nuclear deterrence policy and related conventional defense agenda is also indicative of the military wanting to enjoy the best of both worlds. There were no signs of any desire to reassess defense policy at both strategic and operational levels to reduce conventional forces, especially after Islamabad decided to go overt with its nuclear option. Although there are no examples of reduction of conventional forces by nuclear weapon states, in South Asia any possible reduction is also discouraged because of bureaucratic factors. It would not be an exaggeration to say that the increase in the military's appreciation of the nuclear option developed greatly after it was realized that it could not manage to obtain conventional weapons from the US or it could not possibly eradicate the conventional arms balance *vis-à-vis* India. According to Lt General (Retd.) Hameed Gull there was irritation in military circles regarding the nuclear programme which had resulted in depriving the armed forces of much needed military hardware from the USA. He claimed that during a closed-door annual seminar organized by JCSC on 5 September 1991, there were papers presented by the three services which advocated roll-back of the nuclear pro-gramme.[6] Their concern was that the program blocked the possibility of acquiring weapons from the US. Such argument was, however, ruled out because of the concern for maintaining a certain strategic balance with India. Conventional weapons alone could not guarantee favourable military parity. This did not minimize the significance of conventional weapons for strategic reasons only but for organizational interests as well.

There was also the factor of personal interests in weapons acquisi-tions. During the period under study, particularly from 1990 to 1998, there was much debate on kickbacks related to arms deals. The naval chief, Admiral Mansoor-ul-Haq, was sacked on charges of receiving kickbacks in procurement of three French Agosta-90B submarines. It was, however, interesting that the charges were not pressed any further after his dismissal; instead, two middle-management-level officers were removed. Financial corruption was ingrained in the system owing to the lack of transparency in the policy-making process and the influence of the military's top brass in decision-making.

International arms manufacturers hire most of the senior military personnel after retirement. The arms sellers operate in the client states through their representatives or agents. In Pakistan there are two dis-tinctive categories of people involved with arms trade. There are the 'indentors', who mostly deal with smaller weapons requirements, and the 'representatives', who work for foreign arms manufacturers. These individuals exercise great influence over arms procurement decisions. According to one particular source, the General Staff, Naval Staff and Air Staff requirements are tailored according to the information sup-plied by these agents, and this information is compiled with considera-tion to the business interests of the supplier firms.[7] The representatives with military background exercise influence due to their vast experi-ence of working in the military establishment, the organization of the DG DP, the Ministry of Defense and the organization of the DG MP. These people not only have experience of working in the military bureaucracy but also have a fair number of connections which are exploited for personal benefits. This has proved beneficial for the manu-facturers who are able to circumvent the, otherwise, cumbersome policy-making system. These employees are hired mainly to lobby in the client state for the manufacturer's interests, something I personally experienced during my stint at the naval headquarters. Trying to understand this peculiar behavior, I found out that procurement in all three services was suffering for three reasons: (a) the Ministry of

Defense was not playing a positive role by leaving negotiations and interaction with suppliers to the service headquarters, (b) procurement was handled by non-experts (contract formulation, particularly, is one of the weakest areas of the weapons acquisition exercise), and (c) the absence of a total quality management approach and a transparent process of procurement. There was a lack of separation or division of responsibilities and, in most cases, it was one group in the service headquarters that generated staff requirements for weapons, negotiated with various suppliers, and selected the system. Furthermore, the lack of a project management approach to procurement was astounding. Under the circumstances, it was but natural for these groups to adopt a 'cut and paste' approach for generating staff requirements for weapons, that is, in most cases, to copy the specifications provided by the supplier of the weapon system. The staff requirements produced thus tend to be restricted to one main supplier who manages to succeed in negotiations with a particular service.

The defense decision-making circle is distinguished by the presence of a strong pro-military lobby. This does not mean that influence on the decision-making process was shared equally by the three services: it largely depended upon the significance of individual services in the country's power politics. The Pakistan Army has enjoyed more influence in policy matters than the other two services. The impact of the Air Force's and Navy's strategic planning and thinking on being able to get their plans and acquisitions approved by the government, or have a say in strategic planning, has depended upon their importance for the Army. Since Independence there has been a basic difference in the thinking of the three services. One of the goals of the 1973 White Paper was to narrow this gap but this objective, nevertheless, could never be achieved. The variation in the influence of the three services and the difference in their strategic thinking will be studied in the following sections.

The Army

The Army, relatively speaking, has most power in matters pertaining to arms procurement and general defense decision-making.[8] Military planning actually presents the thinking of Army generals. The political leadership or the other two services have never been equal partners in determining the military strategic priorities. This is a consequence of the Army's comparatively larger size and its involvement in the country's politics. It was no other but the Army chief who enjoyed the dual positions of the head of the service and President of the country.

The Army chief imposing military rule in 1999 used the term 'Chief Executive'. It did not hide the fact, however, that he was in control of the country rather then the President who was retained for the purposes of face saving. The office of the Chief of the Army Staff has emerged as the focal point of power; whether it be in military or political matters, the holder of this position calls the tune. This particular office bearer has always been instrumental in imposing military rule. The other two service chiefs have never been deeply involved in the process.[9] During the period under consideration the Army enjoyed a prominent position in Pakistan's power politics owing to martial law rule from 1977 to 1985. This had a long-lasting effect on the political environment and, as a result, successive political governments, despite the restoration of democracy in 1985, have existed under the psychological and political domination of the Army.

The service's political influence also caused the JCSC to fail in carrying out its prescribed tasks. This institution was founded primarily with the political goal of curtailing the Army's influence in Pakistan's power politics. The plan, nevertheless, could not succeed owing to Prime Minister Zulfiqar Ali Bhutto's persistent dependence on the service's support for his political survival, and subsequently the imposition of martial law in 1977. According to General (Retd.) Muhammad Sharif, who resigned in 1979 as Chairman JCSC in protest against Zia's high-handed attitude, the imposition of martial law soon after the JSCS's creation in 1976 thwarted all efforts at joint military planning and assigning an equal status to the three service chiefs. One of the main reasons for this was that, after the imposition of martial law in 1977, the representation of the three services at the JCSC meetings had become unbalanced. Since the Army chief could not attend the meetings, for by then he had also become the President, the vice chiefs of the Army staff represented him at the forum.[10] This idea was supported by Lt General (Retd.) M. Iqbal Khan as well.[11] The Army deliberately discouraged strengthening the JCSC. Some people involved with the JCSC's planning were of the view that this establishment served a dual purpose of (a) a post office from where the fund allocation is communicated to the services[12] and this allocation follows a rule-of-thumb of an annual ten per cent increase,[13] and (b) serving as a platform for joint discussion where the Army dominates the scene. An effort was made by the political governments to enhance the Committee's influence after 1990. This was done by co-opting its Chairman to assist the civilian decision-makers in influencing arms procurement decisions. The JCSC had asserted itself in supporting the PAF's proposal to acquire the French Mirage 2000-5. Not only was such a procurement plan stalled by trumpeting corruption charges against the decision-makers, the Army subsequently tried to force the new regime in 1997 to scrap the JCSC altogether. The argument was that the Committee was not fulfilling its duty and hence, was a burden on the exchequer. Desirous of resurrecting a strong organization, the civilian government failed to transfer any financial powers to the joint chiefs that would have forced the services headquarters to take the institution more seriously than it did.

The beginning of 1999 saw another threatening change when the chairmanship of the JCSC was given to the Army. Despite it being the Navy's turn to head the organization, the position was passed on to the Army chief, General Pervaiz Musharaf, an action which spoke of the Army's psychological influence over the civilian regime. The naval chief, whose turn it was to become the Chairman JCSC, resigned from his position in protest against the decision eight months before the expiry of his tenure. The government had appointed a junior officer to head the Committee without changing the rules of business. Was it to weaken the military as an institution by pitting one service against the other? This action did not reflect the fair-mindedness of the political government. On the other hand, the Army chief's act of ignoring the principle of seniority by accepting the position of the Chairman exhibited his personal ambitions and a lack of will to concede equal status to the heads of the other two services. It must be noted that both the Naval and Air chiefs were senior to the Army chief.

The larger service's predominant position in power politics and policy-making can also be attributed to certain personalities such as General Zia-ul-Haq. There are reasons to believe that Zia had ambitions of prolonging his rule. This is not the first time that dictators had such plans for sustaining their personal power: the military dictator during the 1960s, Field Marshal Ayub Khan, had a similar ambition. He had declared himself President and had manipulated the elections to stay in power. Zia followed a similar course. In 1984 he had engineered a national referendum in which the question regarding people's choice for the imposition of Islamic rule in the country was equated with the public's preference for Zia's rule. It was to be naturally presumed that if the masses wanted an Islamic system of government they had opted for the military dictator for a period of five years. The results were manipulated.[14] Continuation of his rule had a direct impact on increasing the Army's power. Moreover, this contributed tremendously towards perpetuating the imbalance between the Army and the other services. It was also instrumental in enhancing the value of the Army chief in Pakistan's decision-making process and politics.

This power does not necessarily mean that the Army was not considerate towards the needs of other services. Being the largest service, its equipment and maintenance needs were relatively greater. A look at some partial procurement figures for the three services for FY 1992–93 shows that the Army incurred more expenditure. That year Director-General Procurement (Army) concluded 949 contracts from 1 July 1992 to 30 June 1993 for Rs. 8308.51 million, compared with acquisitions by the Director Procurement (Air Force) who concluded 450 contracts worth Rs. 655.60 million for the same period.[15] This trend cannot be seen in the procurement of major weapons systems. In fact, the Army allowed the PAF to draw maximum benefits from the American military assistance and the Army top brass settled for secondhand/refurbished equipment. Apparently, the GHQ had conformed to the greater strategic requirements of the country. There was a recognized need to strengthen the military's overall defense capability, which could be achieved in a better way by augmenting the Air Force.

The Army's main thrust was to enhance its defensive capability. This was to be achieved through adding an element of air defense. Firepower was increased by adding more tanks and APCs. During the 1980s, considerable attention was paid to the armoured corps. This was done through the procurement of tanks and initiating projects to strengthen the armoured corps.

The Air Force

Compared with the Army, the Air Force is smaller in size, but its requirements for sophisticated technology makes it a capital-intensive service but one that has never had political significance. Nevertheless, this did not stop the PAF's military modernization taking place in the 1980s. A major part of the first American military assistance was spent on procurement of 40 F-16s from the US. A contract for an additional 72 of these aircraft was agreed to be acquired from the second aid package. The primary explanation for this relates to the PAF's significance to the Army and its importance in the country's overall military strategy. The Air Force, because of its superior performance during combat in 1965, and again in 1971, has managed to prove its worth. This strategic significance of the service, which seems to have increased in the 1980s, also resulted in its increased importance as a necessary supporting arm of the Army. This facilitated the procurement of weapons by the PAF and helped get its maximum demands approved. While conceding the need to fulfil the Air Force's requirement for better aircraft, the Army had taken a backseat in getting its own demands met. The US military aid could not sufficiently cater to the overall military modernization requirements of the Pakistan military.

It was thus that the Air Force started to look at the French Mirage 2000-5. Although the other two services had their needs too, it was obvious that two of the three services would have to sacrifice their interests and needs as the PAF high command vied for a bigger share to get the French aircraft. These aircraft had been considered since the end of 1980s. Keeping in view the following three factors, successive civil governments after 1988 had supported the case for the new aircraft:

(a) the publicized threat perception;
(b) the Air Force's inability to procure quality equipment especially from American sources; and
(c) problems encountered by the PAF regarding operation of its F-16s owing to shortage of spares.

The Army was annoyed over this prospective deal that would cost Islamabad approximately $4.2 billion and would naturally diminish any chances of the Army acquiring equipment like tanks, which were needed very badly. It was the President who ultimately blocked the deal. This also reflects the lack of joint strategic planning for defense and a forum where such issues could be resolved. (In the US it is the Congress and in the UK the office of the Chief of Defence Staff which serves this purpose.)

The Navy

The Pakistan Navy is reputed to be the most neglected service of the armed forces. It could not obtain any new major weapon system until the beginning of the 1990s. This is not only because of the lack of political power but its insignificance in the country's defense strategy as well. Military plans in Pakistan are focused on the thinking of the Army generals who have a 'land-locked' approach which makes it difficult for them to appreciate the importance of naval defense. The Navy's top policy-makers feel that the country's decision-making elite is not appreciative of the service's new and multi-dimensional responsibilities of keeping the sea lanes of communication open, asserting its presence in Pakistan's Exclusive Economic Zone (EEZ), and defending against a prospective naval blockade by the Indian Navy. All these objectives call for the allocation of more resources which Army generals do not appear too willing to give. Their approach is based on the national defense strategy which concentrates on land battles with India, particularly on the Kashmir issue. The service's approach on military strategy and threat assessment is different from the Army. Officers in key positions were more concerned about the potential threat of a blockade from a coalition of forces as happened with Iraq during the Gulf crisis. There was a view that the US may want to press Pakistan to give up its nuclear option through blocking its trade and this was the real potential threat, rather than India that the military must be concerned about. This perspective was not shared by the larger service which had a greater role to play in strategic planning. An appreciation of the Navy's argument would possibly get the service more significance and a greater share in resources.

The Army's inconsiderate approach towards the needs of the smallest service did not stop the Navy from vying for more resources. The creation of a separate division of the Marines in the early 1990s for the defense of the Navy's offshore establishments in the presence of SSG force, and an Army contingent dedicated for this purpose, can be interpreted as an effort to strengthen the service's power within the military bureaucracy. Allocation of resources for the service, however, depended to a great extent on the influence and personality of the service chief and his relationship with his counterpart in the Army, or the head of the civilian government. Naval officials are of the view that the service has a better chance of getting their demands approved by the civilian leadership. This, they believe, is because the political leadership is more conscious of the Navy's significance.[16] For instance, it was during Zulfiqar Ali Bhutto's regime that the Navy acquired a separate aviation wing and a few surface ships. Similarly, the service obtained major weapon systems under the political regimes of Benazir Bhutto and Nawaz Sharif. One cannot fail to observe political and personal colours in these procurement decisions. Democratically elected regimes have tended to support the smaller services in order to reduce the Army's influence.

Decisions regarding naval procurement, like acquisitions for the Air Force, are also determined by the personality of the chiefs of services. It is a fact that weapons requirements in the service headquarters are generated in isolation from the actual needs of the end-users – the fighting forces, but by the top echelon of the service[17] (see detailed discussion in Chapters 7 and 8). The hand of the top management, especially the Naval chief, is most visible during the selection process. Officers involved with planning tend to recommend the plans and policies they believe will earn them favour with the top management. The majority of the officers, however, tend to support policies that they feel have a better chance of being sustained, irrespective of the inclination of a particular service chief. Hence, they are not inclined towards policies that may be beneficial for the service but have a lesser chance of being sustained after the change in the top management every three to four years. This is not done through disobeying the service chief but through manipulating information. The middle management of a service that comprises officers of the rank of captain and above have influence in terms of filtering and providing information to the top management that could serve the long-term organizational goals. In weapons procurement and strategic planning there have been instances when information sent to the top management was manipulated or distorted to serve personal or short-term organizational interests. Resultantly, when a head of a service wants to make his organization more efficient there are stumbling blocks created by no one else but the middle management of the service. The most conspicuous example relates to the Navy: the Naval chief, Admiral Fasih Bokhari, tried to make the organization more efficient by re-structuring and downsizing and he encouraged his officers to be more analytical. But most opposition emanated from within the PN.

The Navy's procurement was greatly determined by its role, or lack of a role, in past conflicts with India. Like the Indian Navy, the Pakistani Navy has had to struggle for a role for itself. Its battle for a role, in fact, was proved more difficult mainly because it could not show any significant performance during the two previous wars. The Navy's role was basically that of coastal defense. Security of the EEZ, which the service chose to project in the 1980s, was never pop popd into an equipment procurement policy. Indeed, the Indian or Pakistani navies have never really gone beyond the defense of the SLOCS. With little support from the rest of the military establishment, the service suffered from a dearth of major weapon systems. To strike a reasonable balance against its adversary, the Navy needed to reach a target of twenty to twenty-five surface ships. The surface ships were needed for an offensive capability and for showing a forceful presence but these it could not manage. The other option was to enhance defensive capability with an alternative for some offensive. This was to be carried out through improving the submarine force. The acquisition of the French submarines and the P-3C Orions would, it was hoped, improve the service's ability to launch an offensive. Coastal defense, nevertheless, remained the Navy's main role that was also obvious from certain other acquisitions during the 1990s (see details in Chapter 8).

The nuclear bureaucracy

Although the nuclear bureaucracy in Pakistan is not formally a part of the military establishment, it works closely with it, especially the Army, which is also the primary actor in nuclear decision-making.[18] The nuclear weapons program known as 'Project 706' is under military command.[19] Particularly after Zia's take-over of the country's political control, there was reportedly an increase in the number of Army officers posted at Kahuta, which is Pakistan's main facility for uranium enrichment, and the service of the Army's Engineers Corps was utilized for the program from an early stage. However, Army officials who have some technical knowledge do not generally form part of the higher management of the nuclear bureaucracy. The few military officers at Kahuta basically take care of the administrative work.[20] Dr A.Q. Khan and other scientists working in various organizations enjoy total control over technical matters which explains Dr A.Q. Khan's significance to Pakistan's nuclear programme and in the nuclear bureaucracy. He was responsible for providing Islamabad with technical secrets for building a gas centrifuge for uranium enrichment. Reportedly, he had provided this vital information at a time when the existing corps of Pakistani physicists had tried not to encourage Bhutto to establish a nuclear weapons programme.[21] According to Bhutto's close aide, Kausar Niazi, it was Dr Khan who had offered his services to Bhutto in the early 1970s.[22]

Dr Khan's strategy paid dividends: he won the confidence of Prime Minister, Zulfiqar Ali Bhutto, who had initiated the nuclear project, and earned a special place for himself in the nuclear bureaucracy. It was mainly because of this monoply that Zia did not remove Dr Khan from his position after the military coup in 1977. In fact, an equation was worked out according to which nuclear decision-making rested with the military and technical support being provided by Dr Khan (see Figure 3.1).

Dr Khan, however, represents just one segment of the nuclear bureaucracy. There is also the National Development Complex and the Pakistan Atomic Energy Commission – the two organizations that played a key role in conducting the nuclear tests. The rivalry between these two organizations and Dr Khan's set-up was obvious. A number of scientists who have dedicated their careers to developing the nuclear program feel uncomfortable with the excessive publicity enjoyed by Dr Khan. They also resent the fact that Dr Khan claimed the nuclear development solely as his achievement. It is, none the less, the Army, which exercised informal control of all these institutions, a

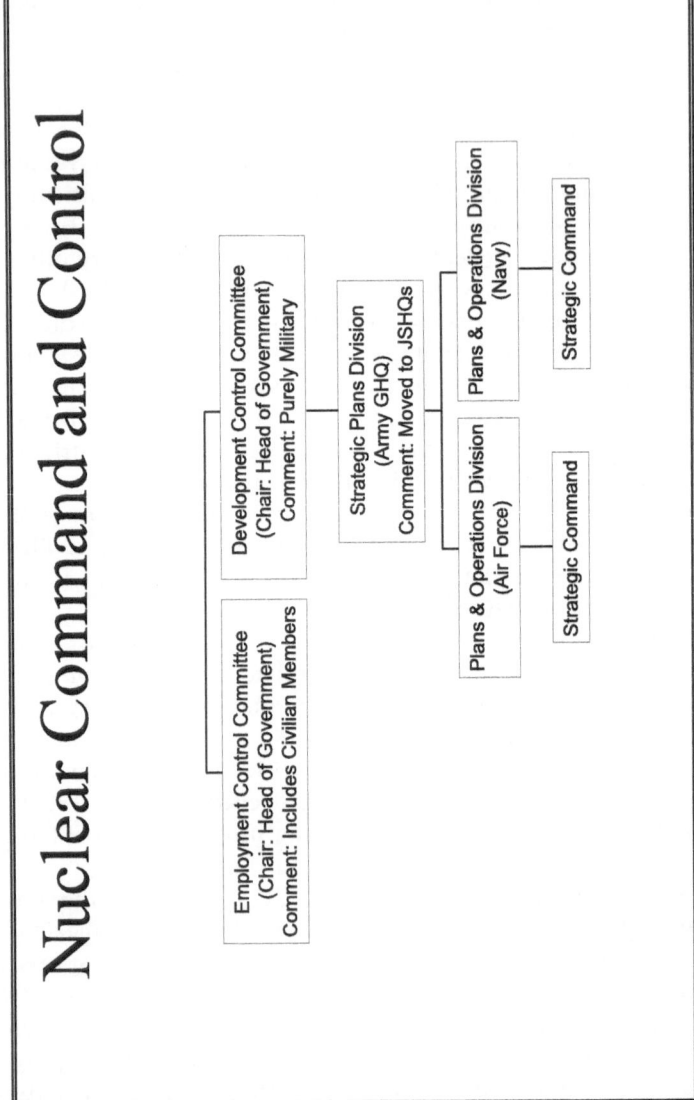

Figure 3.1 Nuclear Command & Control

fact that contained the rivalry factor and stopped it from disturbing the execution of the program. The larger service also discouraged the other two services from an equal participation in nuclear decision-making. The role of the other two services was limited to providing operational assistance in which they were guided by the Army. The Directorate of Strategic Development at the GHQ carried out most of the nuclear operational planning.

In recognition of the nuclear scientific community's know-how, some of its members were included in the Nuclear Command and Control Authority (NCCA) formed in February/March 2000. It is noteworthy that the nuclear scientists alone have the expertise to operate the nuclear weapons, on the direction, of course, of the military.

The civil bureaucracy

The civil bureaucracy involved in defense decision-making refers to three departments: the Ministry of Defense (MoD), Ministry of Finance (MoF), and Ministry of Foreign Affairs (MoFA). These branches serve as junior partners in the military bureaucracy. The common interests, shared memories, and experiences have provided ample understanding of each other's needs. Military regimes, particularly General Zia's government, provided the civil bureaucracy with a sense of stability that was eroded under Zulfiqar Ali Bhutto's government which had sacked 1300 civil servants. Their confidence was eventually rehabilitated under Zia. This was achieved by restoring some of the power and significance of the bureaucrats. The two regimes tended to share their sensitivity towards India. The day-to-day performance, however, depended on an individual department's culture and the importance accorded to it in the overall decision-making process.

The MoD's organizational structure, for example, was fashioned to make it work as the linchpin of the defense organization. Serving and retired military officials were placed in central positions in the Ministry to enable them to control and monitor the work according to the desires of the defense establishment. The two additional secretaries in the defense production division, and three in the defense division, second in command to their respective secretaries, are placed in a position to manipulate the large subordinate civilian workforce. These additional secretaries are posted to the MoD from the three services of the armed forces. This, it is claimed, was done to ease the problems that civilian bureaucrats might encounter in understanding the military's strategic needs.[23] It was obviously presumed that civilian officials of the Ministry were not capable of handling military affairs on their own. The truth is that such logic is used to accommodate senior military officers who are inducted in the MoD prior to their retirement basically to watch over the interests of the armed forces. This they do successfully despite the presence of civilian bosses, the federal secretaries, who control the two departments of the MoD. Even these civilian officers were replaced in 1997 with officers from the Army, which was a clear departure from Bhutto's days when he had sought the services of people from the private sector like Masood Hassan who was appointed Secretary for Defense and served in the position until the end of the 1970s. The move can be considered as one of the many tactics of the Sharif Government to appease the military. In a country where the political and decision-making culture has a military orientation, it would not be difficult for a handful of military officers to control a larger number of civilians.

The peculiar placement of military officers serves personal interests as well. Serving military personnel, posted in MoD almost at the end of their career or after retirement get a chance to create future opportunities by working in the Ministry. There is a tendency to support decisions that can later help them build alternative career opportunities, for instance, as agents for arms manufacturers, or in some cases, accumulate assets to see them comfortably through their retirement. It was to check this abuse of authority that the government in 1991 had considered passing a law which would bar retired defense services personnel from accepting jobs with international arms manufacturers. Such a law, nevertheless, was never implemented.[24]

The other ministry linked with the arms procurement process is Foreign Affairs, which assists the government in finding sources of supply for weapons. This has been termed as its sole contribution to the process.[25] The Ministry is not proactive in the procurement process. Speaking from personal experience, I believe military officers have reservations in consulting MoFA on a regular basis on defense-related issues. Others, such as the former Foreign Secretary, Shehryar Khan, argued that this Ministry's importance in arms procurement decision-making has varied with the heads of government. He claimed that General Zia used to seek the advice of the Foreign Office about possible channels of arms supply, unlike the policy pursued by the political governments.[26] What he failed to mention, however, is the fact that being the President and a military dictator Zia had more control over the government. This gave him a level of confidence which no political government after 1985 enjoyed in seeking views of other ministries. The fact is, the Foreign Office is rarely consulted in matters pertaining to arms procurement or military planning. The Kargil operation was a glaring example of the communication gap between the Foreign Office and the military establishment.

From the arms procurement perspective there are two broad dimensions of Pakistan's foreign policy: (a) it is predominantly India-oriented, and (b) focused on two major sources, the US and China. The concentration on these two countries resulted in the Foreign Office failing to develop business relations and improve diplomatic ties with other countries, especially the former Soviet bloc states. Foreign policy-making was found to be *ad hoc*, hence not providing the dividends to Islamabad in its battle against India, or in finding sources of supply other than Washington. Poland's refusal to provide spares to Ukraine for the manufacture of T-80U tanks for the Pakistan Army on diplomatic grounds can be deemed as one such example. Warsaw's position was that it was unwilling to supply weapon systems and spares to a country hostile to India. It is interesting that this position was adopted after Islamabad selected the Ukrainian tanks instead of Polish. This does not mean that the policy towards China or the US was perfect. Beijing is considered to be a vital source of help to Islamabad in countering the Indian threat but their bilateral military links, nevertheless, underwent a qualitative change in the 1980s and narrowed down to mainly an arms transfer link (see Chapter 5). Even in this case the Foreign Office's performance was not very impressive.

As for the US, Pakistan's relations with Washington during this period were marked with greater fluctuation than those with China. This was primarily due to the changes in American foreign and defense policies which caused the convergence, and, later, divergence of views between the two allies. Equally responsible were the views of various influential people and groups from the decision-making elite. Three distinctive groups were identified.[27] The first was categorized as the 'independence' group, whose members attach more significance to economic prowess playing a greater role in determining bilateral relations in the post-Cold War environment. Close to this is the second type reputed as the 'Muslim' group that advocates political and security alignments based on religious ideology. People from these two categories have a confrontational attitude with the US regarding certain issues like nuclear proliferation. The latter school of thought believes that Pakistan must project itself as one of the leaders of the Islamic world – a position that can be enjoyed on the basis of the country's nuclear capability and a relative control over Central Asian politics. What makes these views important is that they are prevalent in the

Army as well. The last kind is known as the 'surrender' group, which supports reliance on the US.

The third important department, closely involved with arms procurement, is the Ministry of Finance. It controls the purse strings of the defense establishment. Military personnel can often be heard complaining about the stringent attitude of the finance officials, but this does not indicate that the Ministry can affect defense decision-making or enjoy the power to override decisions taken by the military. Their supposed reluctance is related to the inertia of the bureaucratic environment and financial constraints of the country. One can appreciate the problems of the Ministry considering situations like the summer of 1993, when the national foreign exchange reserves were at a record low: $200 million only. The finance people faced a 'push–pull' situation with the loan-giving agencies/governments pushing it to invest in other sectors and to reduce the deficit.

There are also demands of the social sector that are generally ignored in Pakistan. The most overbearing pressure is from the military to provide funds for the maintenance of existing infrastructure and fresh acquisitions. Despite serious resource-limitation the Ministry is not able to dictate its terms to the military establishment. All it can do is delay decisions, which in fact proves to be negative. If a service is determined to acquire a certain weapon type, it eventually succeeds by pushing it through the DCC. The impact of the delay is an increase in the weapon price. The comparative influence of the military can also be observed by looking at the audit reports of the defense organization. They show that the MoD and the armed forces do not attach much importance to the objections raised by finance officials regarding misappropriation of resources.[28] Officials of the department of the Auditor-General of Pakistan, part of the Ministry of Finance, encounter similar problems while auditing arms procurement expenditure. They are usually not allowed to review certain documents which are necessary to carry out a complete audit of these accounts.

The most unfortunate fact is that non-experts are running the Ministry. These are officers inducted into the Ministry through a qualification exam that does not specify knowledge of finance and economics as a prerequisite. The bias of the political leadership for the defense sector further hampers good financial management. This factor alone has played a role in not making the military appreciate the sensitivity of sound financial management. The armed forces tend to shy away from comprehending the financial aspect of procurement and defense buildup, or its implications on other sectors and the national growth. It is believed that foreign relations and finance are the issues that would automatically be taken care of by the government. Such a notion resulted in frustration and fiascos on a number of occasions and the finance people did not contribute much in educating military personnel on the subject. During the financial crisis that ensued after the nuclear tests in 1998, the Finance Ministry and the government did not actively apprise the military establishment of the dire economic constraints. As a result, the element of frustration grew when the government could not provide funding for various defense projects requested by the services. This non-sharing of information would not have helped the military learn about the new financial environment in which the state needed to cater to development and other needs, rather than struggle to make 'both ends meet.'

The President, Prime Minister and the parliament

During 1979–99 the President emerged as one of the key actors in defense/arms procurement policy-making. This was due to three factors. First, the constitutional power acquired by President Zia-ul-Haq through the eighth amendment empowered him to dismiss a government and appoint the Chairman JCSC and the three service chiefs. This legacy was passed on to the Presidents who followed. Moreover, this authority was exercised frequently, especially in 1989–90, when a controversy surged between the President and the Prime Minister over the appointment of Admiral Sirohey as Chairman JCSC; it was the former's will that prevailed. Second, was the influence of various personalities such as General Zia, who strengthened the office of the President. The military dictator had installed himself in 1977 as the Chief Martial Law Administrator (CMLA) and later decreed himself President, resulting in the strengthening of his position. He continued to enjoy this status even after the restoration of democracy in 1985. After his death in August 1988 it was his close aide, Ghulam Ishaq Khan, who became President.

Third, in the absence of a constitutional position, the military resorted to co-opting the President as a junior partner in defense decision-making and formulation of policies that affected the armed forces. The Army in particular was conscious of the fact that by the end of the 1980s, and after the end of the Cold War, a military dictatorship was no longer acceptable to its Western ally, the US. It was, however, ensured that the defense establishment's interests would be taken care of. Ghulam Ishaq Khan was allowed to take over as President after General Zia's death because of the rapport he had developed with the military.

President Khan also provided the military with a cushion against the then newly elected Prime Minister, Benazir Bhutto. The Army top brass was skeptical of her and one government of Mian Nawaz Sharif trusted Ishaq to guard their interests by keeping her in check.[29] This point of view was reinforced by repeatedly sending two governments of Benazir Bhutto and one government of Mian Nawaz Sharif home. The President was there to guard and cater for the military's long-term political interest by keeping in check any ambitious political regime and endorsing plans that are supported by the armed forces. For instance, Ghulam Ishaq Khan agreed to give extraordinary powers to the Army in Sindh.

The parliament elected in 1997 repealed article 58 (2B) that had given extraordinary powers to the President. This resulted in a situation in 1999 where the Army had no constitutional option to remove the political government but through direct interference. The President, as mentioned earlier, in the past played this role. Nawaz Sharif was accused of conspiring to destroy the institution of the military. Notwithstanding the lack of Sharif's political acumen and the mistakes he made, he had threatened the military's core interests, which led to his removal. Administrative control of the armed forces and general military planning are areas where the armed forces do not allow any interference. Prime Minister Junejo, it must be remembered, was removed in 1988 because he had tried to take a course opposed to the Army. His efforts to probe into the 'Ojhri' camp disaster and introduce austerity measures in the defense establishment had backfired. Benazir Bhutto's government elected in 1988 had a similar fate. The Army manipulated her removal in 1990. Despite her knowledge of this information she remained silent which was due to her fear of the Army's possible retaliation.[30] This fear was instilled in the hearts of all governments owing to the Army's manipulations. The technique was to encourage a negative competition between the ruling party and opposition. This had two obvious effects: first, the ruling party, being conscious of the Army's power to destabilize the regime, conceded to its demands. Second, it weakened the elected parliament. In any case, the Pakistani parliament was never strong enough to impose its will on the military. Traditionally, the elected representatives do not enjoy the power to debate on defense expenditure. In the national budget defense spending is categorized as 'charged' expenditure on which a public debate cannot take place. This automatically reduces any chance of the parliament playing a vital role in defense decision-making. However, the responsibility for this state of affairs did not lie entirely with the military. There are three basic reasons for this: (a) the elected body of the country mainly comprises members of the land-owning class who have traditionally operated in collusion with the military and civil bureaucracies;[31] (b) the educational standard of the parliamentarians is generally low, which adds to their general inability to question arms procurement or any other defense decisions; and (c) the political leadership, for reasons of personal ambition for power, deliberately supported the military and its demands. This presented the politicians' belief in the armed forces being the key actor in the country's power politics. For instance, in 1997 when Prime Minister Sharif started to talk about reducing military expenditure it was Benazir Bhutto who became vocal about the military's threat perception and requirements. She was obviously trying to convince the armed forces that were instrumental in her removal of her good intentions towards them. A western model of democracy is unacceptable to a military that does not sufficiently trust political leadership to guard the organizational interests of the armed forces.

Indirect actors

There is yet another category of actors that play an indirect, albeit, important role in building the influence of the military. Not all of these players benefit directly from military modernization or defense planning. Their conscious or unconscious moves, however, help in strengthening the military as an institution. There are three types of actors in this category: (a) military intelligence agencies, (b) religious fundamentalist groups, and (c) the media.

Military intelligence agencies

Pakistan's military intelligence agencies have played a significant role in its power politics, formulation of threat perception and running independent defense policy particularly in Central Asia. Each service of Pakistan's armed forces has its own intelligence branch known as Military Intelligence (Army), Air Intelligence and Naval Intelligence. The most significant, nevertheless, is the Inter-Services Intelligence popularly known as the ISI. Since its involvement in the Afghan crisis when this agency was given a free hand in running operations, it emerged as an influential partner within the military bureaucracy. As a result, Islamabad's Afghan policy after the Soviet withdrawal from Afghanistan fell totally within ISI's ambit. This was true in the case of operations in Indian held Kashmir as well, which was primarily an Army-ISI affair.

What further strengthened the institution was its involvement in domestic politics being instrumental in creating the *Islami Jamhoori Ittehad* (IJI) and *Mohajir Qaumi* Movement (MQM) to counter Bhutto's PPP. This was admitted by its former Director-General, Lt General (Retd.) Hameed Gull, during an interview with the author.[32] Again in 1990 the ISI was used to destabilize Benazir Bhutto's government, a strategy that was adopted to stop democratic governments and institutions from strengthening themselves – a factor that would be detrimental to the Army's political interests. One of the common bonds between the ISI and the Army was financial mismanagement that became very obvious during the 'Mehran' Bank scandal pertaining to monetary bribes given to the agency to destabilize Bhutto's government. In this case, the chief executive of the bank, Yunis Habib, admitted to having provided General Mirza Aslam Baig, then the Army chief, and ISI with fourteen million rupees that were used to manipulate the 1990 elections.[33] General Baig did not deny the charges: his only observation was that there was already precedence of the ISI taking money from private entrepreneurs.[34]

The Army controls the ISI and, in spite of the ISI chief being appointed by the Prime Minister, the core of its personnel is drawn from the Army. This provides the Army chief with substantial leverage in using the institution to serve the greater organizational interest of the armed forces.

Religious fundamentalists

The term refers to all ideological and religious groups in Pakistani society whose socio-political manifesto is based on the desire to transform the country into a religious fundamentalist state. The most prominent of these groups is the *Jama'at-i-Islami*, which gained importance during the 1980s because of General Zia's Islamic orientation. It was due to his support that the *Jama'at* managed to make inroads within the military and have an impact on society. It is no secret that the *Jama'at* and ISI worked closely during the Afghan operation, and even after the crisis on other fronts as well.[35] This religious extremist party has an organized force to conduct its operations in Kashmir consisting of separate political, military and publicity wings. Its support was considered vital by the military decision-makers due to the financial and human support which the *Jama'at* was capable of obtaining from other Muslim states on ideological grounds. The party's leadership projects an aggressive stance towards India, particularly on the issue of disputed territory, one of the cornerstones of the party's linkage with the Army. Its politics focused on Kashmir in order to muster public support, a strategy adopted by most governments and the armed forces in order to win popularity. Such an argument has been based on the knowledge of the history of *Jama'at's* stand on the Kashmir issue, which previously had been fairly independent of the military's views. This attitude changed later with the party willing to cash in upon the growing emotional significance of the Kashmir issue to win support of the masses and the military. The significance of this transformation lay in the party's ability to influence the public emotionally on the subject. This fundamentalist approach to conducting relations with other states, and the publicity of this peculiar viewpoint, was beneficial for the military as well. This was because the *Jama'at* and other fundamentalist parties were adept at invoking the religious sentiments of the populace on the Kashmir and nuclear proliferation issues.

The media

Pakistani media has always been predominantly government oriented. It does not have independence in projecting views other than that of the government. This is the hallmark of electronic media, mainly state owned, and printed media as well. Although the government does not own the print media, there is a tendency to present the state's point of view, especially on defense related issues. One explanation for this is that discussion on matters relating to national security is considered taboo. In Pakistan the presentation of radical views on such issues continues to be difficult. Despite the fact that the media exhibited far more openness in discussing Islamabad's nuclear proliferation policy than the Indian press, members of the press do not have a high competence or knowledge to discuss issues pertaining to the technical aspects of military modernization. A look at major newspapers shows that they tend to present an official point of view. Moreover, media has played a vital role in building the military's positive image by presenting views in the interest of the armed forces. For example, the Pakistani media contributed immensely towards giving an emotional dimension to both the nuclear and Kashmir issues and the need for maintaining a military posture. This factor has made it impossible for any regime to change existing policy in these areas.

The printed media has played a significant role as well. In the absence of a forum where problems relating to inter-service rivalry could be addressed, this part was played by the media. Starting from the end of the 1980s, national newspapers became a vehicle for the two smaller services to express their grievances, especially pertaining to their arms requirements and overall strategic needs, or talk about their threat perception. These publications are important for catching the attention of the public and the top civilian policy-makers who other-wise had no access to any independent source of information. Sometimes this had a direct impact on decision-making; for example, the Mirage 2000-5 deal was stalled mainly through negative publicity in the newspapers. Although this development denotes a peculiar strengthening of the democratically elected leadership, this cannot be construed as the media's enhanced ability to question the military or lessen its image. The continuation of a positive image was the mili-tary's requirement for strengthening its position in the country's power politics, a status that has been used to get its demands approved by both the civil regimes and the general public. The use of the armed forces' influence on decision-making was accompanied by a lack of transparency in the policy-making process. Not only did that make certain decisions debatable, it also added to the economic burden of the country without bringing about a force multiplying effect. The cost factor is a vital angle that will be analysed in the next chapter.

4
The Cost of Military Buildup

For Islamabad the cost of arms procurement and military buildup has been extremely high. The increase in threat perception during the period increased the pressure on Islamabad to invest in the defense sector at the cost of socioeconomic development. During a considerable part of these twenty-one years, weapons modernization was partly financed by American military assistance but this did not decrease the defense burden on the national economy and, in any case, American military aid was interrupted in 1990 after Washington imposed an embargo. What added to the burden was the discontinuation of American economic assistance as well. All this happened at a time when Pakistan's economic conditions had become bleak and depressing.

Defense: the economic burden

One of the prominent features of Pakistan's national budget is the high concentration of spending on debt servicing and defense. Since the country's birth in 1947, the national budget has always indicated the government's bias for national security. Policy-makers, who preferred to concentrate on military security, were always nervous about the military capability gap between Pakistan and India and this resulted in their choice to invest funds in strengthening military security. The priority attached to territorial security has cushioned defense spending from all other needs.[1] There was also the belief that the military and its spending were important contributory factors towards socioeconomic growth and industrial development. In the 1970s and 1980s the growth in GDP was linked with the increase in investment in services such as defense and public administration.[2] Although this argument is debatable considering the impact on social growth. High defense spending was a trend followed throughout the twenty-one years under study (see Table 4.1). This was also the period that procurement budget gradually increased (see Table 4.2).

Table 4.2 does not give any details of the equipment that was procured during this period. The author has tried to put the information together on major arms procurement and make some sketchy calculations about the burden of weapons modernization, shown in Table 4.3. The increase in the financial burden linked with these acquisitions is obvious. A major portion of the burden was for a service that had no strategic value in the country's war fighting plans. The two procurements for the Navy did not signify any change in the thinking of the Army regarding the necessity to safeguard the EEZ and SLOCS. Reportedly, the deal was opposed under Zia's rule.[3] The naval chief, Mansoor-ul-Haq, had managed these acquisitions owing to his personal contacts with the top political leadership and lobbying within the armed forces. This was achieved without a real shift in military thinking, a fact obvious from the division of resources within the armed forces.

According to Figure 4.1 the Army received the largest chunk of the resource pie, a major part of the budget of each service being spent on the pay and allowances of officials. The division of the Army's budget given in Figure 4.2 is an example of evaluating the spending pattern of the services.[4] The total number of workers in the military was the main reason for the high personnel and maintenance cost. It would, therefore, be difficult for Islamabad to reduce its defense spending without

Table 4.1   Pakistan's Official Defense Budget for FY 1971–72 to 1998–99

FY	Defense budget	FY	Defense budget
1977–78	9 674.00	1988–89	51 053.00
1978–79	10 302.00	1989–90	58 708.00
1979–80	12 655.00	1990–91	64 623.00
1980–81	15 300.00	1991–92	75 751.00
1981–82	18 631.00	1992–93	87 461.00
1982–83	23 224.00	1993–94	91 776.00
1983–84	26 798.00	1994–95	100 221.00
1984–85	31 866.00	1995–96	119 658.00
1985–86	35 606.00	1996–97	127 400.00
1986–87	41 335.00	1997–98	133 800.00
1987–88	47 015.00	1998–99	145 000.00

Note: All figures are in rupees million.
Source: Economic Survey of Pakistan.

Table 4.2 India and Pakistan Arms Imports, 1985–95

Year	India			Pakistan		
	Current	*Constant 1995*	*Arms imports % of total imports*	*Current*	*Constant 1995*	*Arms imports % of total imports*
1985	$2 600	$3 563	16.30	$470	$644	8.00
1986	$3 200	$4 271	20.80	$330	$440	6.10
1987	$3 000	$3 883	18.00	$340	$440	5.80
1988	$3 100	$3 870	16.20	$460	$574	7.00
1989	$3 000	$3 595	14.60	$550	$659	7.70
1990	$1 800	$2 069	7.60	$925	$1 063	12.50
1991	$925	$1 022	4.50	$220	$243	2.60
1992	$650	$699	2.80	$450	$484	4.80
1993	$270	$283	1.20	$550	$577	5.80
1994	$230	$236	0.90	$290	$297	3.30
1995	$410	$410	1.20	$480	$480	4.20

Note: All figures are in US $ million.
Source: US ACDA, 1996

Table 4.3   The Weapons Procurement Burden, FY 1992–93 to 1998–99

FY	Service	Source	Weapon type	Quantity	Procurement category	Cost
1992–93	Navy	UK	Frigates	6	Secondhand	$150
1994–95	Navy	France	Minehunters	3	New	$350
1994–95	Navy	France	Submarines	3	New	$1 300
1996–97	Army	Ukraine	Tanks	320	New	$650
1996–97	Air Force	China	Jet Trainer	6	New	$20
1996–97	Air Force	PAC, Kamra	Aircraft	30	Overhaul	$116
1996–97	Air Force	France	Aircraft	40	Overhaul	$118
Total:						$2 704

Note: All figures are in US $ million.

decreasing manpower. Personnel and maintenance costs, it must be remembered, denote fixed costs. Major weapons acquisitions are an additional cost incurred by equipping the large workforce. It also makes the total burden more than one could imagine.

There are two important dimensions related with this assessment that need explaining. First, one of the many reasons for continued high defense spending was the high percentage of wastage of resources, which grew as a result of oversight and encouragement of the top managers at the highest decision-making level in MoD and the armed forces. In 1998–99 Islamabad was wasting approximately 30 per cent of its defense budget. This was because of certain indigenous defense production projects, excess number of employees, wasteful imports such as cars and other luxury items for senior officers, duplication of activities, and corruption.[5] Pilferage of financial resources dates back to the start of the Zia era when the military dictator encouraged certain practices to please his senior officers. For instance, in 1977 he allowed the practice of maintaining special slush funds by the corps commanders who were at liberty to use public funds under this particular head at their discretion. It was after a long time and in the face of great resistance that the public sector auditors were allowed finally to audit the accounts related to the expenditure of such special funds. The audit exercise was limited in nature, however.

Second, excess expenditure was met from extra-budgetary sources. Funds were diverted from the profits of organizations established for the welfare of retired military personnel. The financial diversion was made mostly to meet personnel cost. The four organisations: the Fauji Foundation (est. 1960), Army Welfare Trust (AWT) (est. 1979), Shaheen

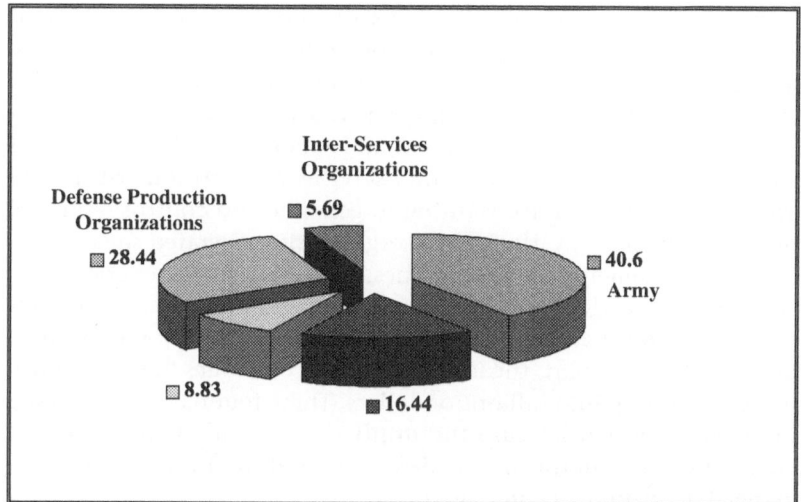

Figure 4.1   Division of the Defense Budget

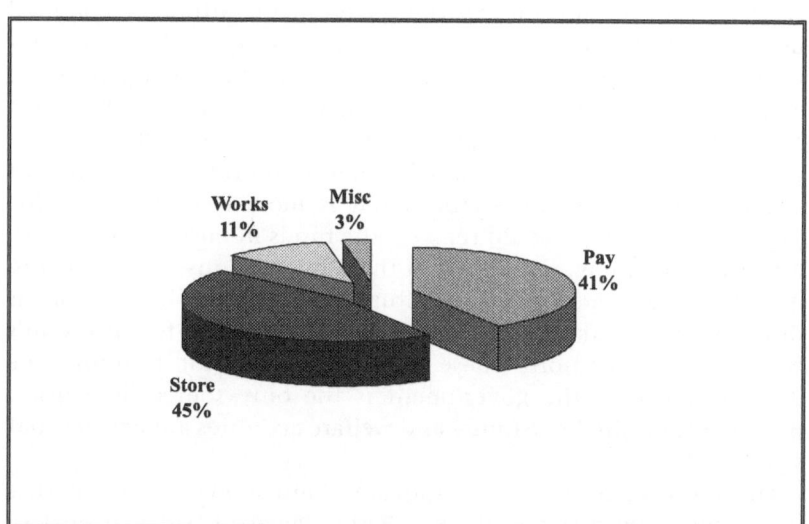

Figure 4.2   Division of the Army Budget

Foundation (est. 1977), and Bahria Foundation (est. 1981) were created with the objective of contributing to the pension fund and welfare of retired military personnel. The first to be established was the Fauji Foundation, meant for the personnel of all three services, but the Army predominantly controls it. The formal control of the Foundation rests with the MoD. It was for unexplained reasons that the AWT was established for the same objective as the Fauji Foundation. This development denoted nothing else but the inter-services competition within the armed forces, each service wanting to generate and control resources for its own benefit. By 1997, these organization operated sugar mills, fertilizer and cement factories, airlines, security agencies, courier agencies, banks, petrol/gas stations, estate agencies, and various other projects. The AWT had the largest number of projects,[6] and is thus, according to one report, the largest group of companies in the country, its assets running into billions of dollars. These foundations increased the defense burden because the utilities and some of the overhead costs were met from the annual defense allocation. In this respect the financial dividend was eliminated.

These foundations are a burden on the economy in three respects: (a) since these are categorized as welfare institutions for the armed forces, they do not pay taxes for their corporate ventures resulting in considerable lost revenue to the government (although marginal taxes were imposed in 1992 on these organizations, the Army controlled foundation was levied at a lesser rate); (b) these organizations employ serving military officials who are paid from defense budgets,[7] and (c) they represent monopolistic tendencies. In the last case, these organizations manage to acquire business opportunities that are not available to the private sector. What is more is that, unlike the private sector which would reinvest the funds in business leading to growth of economy, the profits of these foundations are diverted to meet the excess personnel expenditure of the services. It, therefore, is difficult to calculate the actual defense burden for Pakistan's economy. Furthermore, these institutions are not fulfilling the desired objective: the government is the only source of pension payment for retired personnel and welfare activities are neither that significant.

The military's corporate activities are a burden on an economy that is in acutely poor shape. In the 1970s Islamabad had managed to offset the negative balance of its trade account and keep its deficit spending at a low level as a result of foreign remittances, particularly from Pakistanis working in the Middle East. It reached its peak in FY

1982–83 with $2886 million after which it started to decline reaching $1461 in 1995–96. Islamabad did not change its economic policy to deal with this reality mainly because the gap was filled by American economic and military assistance in the 1980s. The situation continued until 1988 when the multilateral and bilateral economic aid donors refused to extend the favorable treatment to Pakistan that it had enjoyed. Successive regimes found it difficult to finance their budget deficit that increased at a fast pace. They resorted to negotiating commercial loans with the result that the debt burden ballooned to over $31 billion. This external debt was in addition to the Rs. 1100 billion in domestic borrowing.[8] Most of the money was borrowed to meet the government's deficit spending (see Table 4.4).

Hence, high defense expenditure was a continued drain on national resources, although certain academics such as Kupchan disagree with this view. According to him, military expenditure does not cause inflation or the deficit[9] but his notion is not based on an analysis of the spending behavior of developing countries such as Pakistan. At the same time, one can agree in part with his analysis because financial mismanagement and corruption is rife in all Pakistan society and it would be unfair to accuse the defense sector for causing the entire financial burden on the exchequer. However, elimination of corruption and reduction of the defense burden may be recommended as a two-pronged strategy for reducing the deficit.

Table 4.4 Pakistan's Annual Deficit

FY	Deficit	Percentage of GDP
1982–83	25 654.00	7.1
1983–84	25 147.00	6.0
1984–85	36 777.00	7.8
1985–86	41 644.00	8.1
1986–87	46 710.00	8.2
1987–88	57 563.00	8.5
1988–89	56 879.00	7.4
1989–90	56 060.00	6.5
1990–91	89 193.00	8.7
1991–92	89 970.00	7.4
1992–93	107 525.00	7.9
1993–94	92 179.00	5.9
1994–95	103 405.00	5.5

Note: All figures are in Rs. million.
Source: Economic Survey of Pakistan.

The year 1998 was the beginning of a tough financial ordeal for the country. The ill-planned economic policies of past governments, especially Sharif's, led to an almost total financial collapse. Following are a few reasons that led the country to almost financial collapse:

- the absence of a transparent and healthy trade policy;
- infrastructure problems, such as load-shedding of electricity;
- the battle between the government and the International Power Providers (IPPs) (as Sharif had refused to honor the commitment of the previous government); and
- corruption of the tax department and the government's inability to meet revenue generation targets, causing problems with the International Monetary Fund.

Moreover, the treatment of the IPPs resulted in the reduction of direct foreign investment. It was in early 1999 that Islamabad defaulted on a $25 million loan from the Islamic Development Bank.

In the middle of the crisis the Prime Minister did not cease to pursue questionable huge projects such as the construction of motorways, the building of a vast thirty-one-gate airport at Lahore and the Prime Minister's housing scheme. All these were expensive projects. Nor did he send the appropriate signals to the military to tighten its belt. The armed forces planned a procurement of about $10 billion to be carried out over a period of seven to ten years. New Delhi's announcement of spending approximately Indian Rs. 700 billion on building a nuclear arsenal was bound to solicit a response from the Pakistani military yet how it would manage to generate sufficient resources to do that was uncertain. One way would be to divert more resources from the social sector. The Federal Finance Minister, Ishaq Dar, is on record as having said that more taxes would have to be levied to provide for national security. Given the Pakistani taxation system's infrastructure restraints, it is likely that the burden would be shifted to existing taxpayers who appear already to be overburdened by indirect taxes. A diversion of resources from socioeconomic development, on the other hand, would quadruple the cost of military security.

Social cost

The overall impact of Islamabad's borrowing and defense spending was not limited. The negative effects could be observed in the form of the retarded growth of social and other sectors as well. Indeed, in

Pakistan's case one does not find the traditional 'guns-for-butter' approach; a factor linked with the peculiar perception of the decision-makers. Allocations were made without taking the financial capacity of the economy or the need for social development into consideration.

Pakistan's social and human development indicators are quite poor. According to the human development report on South Asia at least 28 million people live below the poverty line; two-thirds of its adult population is illiterate; basic health facilities are available to only half the population; maternal mortality rate is very high at 340 per 100 000; one-fourth of new-born babies are underweight and malnourished; it has the highest population growth rate in the region at 3.6 per cent.[10] In human development terms, the country laggs behind other major regional states such as India and Sri Lanka. This was despite the *per capita* increase in income by 231 per cent for a period from 1970–93. This was claimed to be the highest rate of increase in South Asia. A comparatively positive *per capita* income growth rate did not help in alleviating the deprivation of the poor. The bad communication network, lack of easy accessibility to social services, and the domination of a feudal life-style upheld by landowners, business class, bureaucrats and other affluent people are some of the factors that add to the ugly picture of the mass of suffering humanity in the country. The government's resource allocation policies were also biased against social development and in favor of defense, a fact obvious from the figures given in Table 4.5. This kind of unequal distribution of resources led to a situation where the number of poor people increased from 19 million in 1960 to 42 million in 1995.[11]

The official picture of socioeconomic and human resource development, of course, was always much more rosy. There are people who argue against the official claim of 36 per cent literacy rate.[12] When looking at the state of the poor in the country one wonders what the figures actually mean in the context of the suffering masses. How usable is the quantified literacy or how efficient is the presently functioning health care system? Pakistan lags behind in science and technology and related education – factors that have been stated as crucial in the development process by some of the theories of modernization. The situation will remain stagnant without any chance of improvement unless there is a substantial commitment of effort and planning and resources by Islamabad. The government has to ask itself the question, how long will it neglect human development especially at the cost of military security?

Table 4.5   Pakistan: Defense *versus* Development

FY	Health*	Education+	Defense
1981–82	0.6	1.4	5.7
1982–83	0.6	1.5	6.4
1983–84	0.6	1.6	6.4
1984–85	0.7	1.8	6.7
1985–86	0.7	2.3	6.9
1986–87	0.8	2.4	7.2
1987–88	1.0	2.4	7.0
1988–89	1.0	2.1	6.6
1989–90	0.9	2.2	6.8
1990–91	0.8	2.1	6.3
1991–92	0.7	2.2	6.3
1992–93	0.7	2.4	6.0
1993–94	0.7	2.2	5.6
1994–95	0.7	2.4	5.5
1995–96	0.8	2.4	6.2
1996–97	0.8	2.5	6.5
1997–98	0.7	2.3	6.9
1998–99	0.7	2.2	7.1

Note: * Expenditure on Health and Education is percentage of GNP.
+ Expenditure on Defense is percentage of GDP.
Source: Economic Survey of Pakistan.

There have been occasions when questions were raised about the defense burden and its impact on the economic performance of the state. During the early 1980s, particularly under Prime Minister Muhammad Khan Junejo's rule, there was a demand to freeze military related spending.[13] Again, it was during Junejo's regime that the Chairman of the Public Accounts Committee initiated a move to establish a separate audit wing to carry out the financial evaluation of military purchases. The office of the Director-General Audit (Defense Purchase) was established towards the end of the 1980s, but by then the Junejo government and the prime mover of the proposal were not in existence to see to the efficient functioning of the organization. The organization thus remained without producing desired results. The civilian governments continued to be deprived of the authority to ensure efficient use of resources for military modernization.

Islamabad continued to resist any move to reduce defense spending until 1988 when the Ministry of Finance under Mahbub-ul-Haq decided to make some reductions. It was nothing significant, however. A Rs. 2.5-billion cut, that would also affect the defense budget, was actually part of the across-the-board reduction in the government's administrative expenses. Zia, who was President at the time, opposed any suggestion regarding cuts in defense spending by using the typical argument related to the country's threat perception. According to him: 'How can you fight a nuclear submarine or an aircraft carrier with a bamboo stick? We have to match sword with sword, tank with tank, and destroyer with destroyer. The situation demands that national defense be bolstered and Pakistan cannot afford any cut or freeze in defense expenditure, since you cannot freeze the threat to Pakistan's security.[14]

The new administration of Nawaz Sharif, empowered in 1997, proved equally ineffective in making the defense sector more account- able, or reducing military expenditure. He had hoped to improve econ- omic performance to a degree where there would be no need for cutting down the defense burden. His pronouncement to match Pakistan's economic growth with the Asian Tigers could not material- ize and this was a promise he utterly failed to fulfil. In six months, from April to September 1998, national exports dropped by 18 per cent. They crashed down by another 20 per cent after June 1999. The tax collections and export targets were not met, this despite the devalu- ation of rupee against the US dollar by 17 per cent in almost six months in 1998. Since the nuclear tests, two rates for foreign exchange conversion prevailed – official and black market. For local consumers and the business community it was the black market rate that was important. This was because with a national foreign exchange balance dwindled to $1.2 billion the government issued the order to block any legal foreign exchange transactions by nationals. The business commu- nity could only remit foreign currency after having purchased it from the black market. A scheme to pay back loans through making appeals to the nationals in the country and the expatriates did not pay off either. This was the strategy adopted by Islamabad to reduce national debt and improve economic growth after tremendous pressure from multilateral economic aid donors.

The economic aid donors asked Islamabad to reduce its deficit spend- ing through decreasing its non-development expenditure, the most significant being defense. Successive governments attempted to quieten the donors through arguing about the threat perception, yet economic aid donors consider national security as a purely domestic matter. Pakistan's major economic aid donors followed this approach. The main focus for Islamabad was to reduce its deficit spending without necessarily decreasing the defense burden. It was precisely because national security interests were held so sacrosanct that the IMF or any other lender did not question Islamabad's off-budget financing of its military expenditure.[15] The Fund would ignore Pakistan's defense burden if Islamabad could improve its overall financial position through carrying out certain structural adjustments for which the IMF committed $1.37 billion in 1994 to be used over a period of three years. These adjustments were aimed at increasing Pakistan's resource generation capacity and would be achieved through tax reforms and other similar measures.

Neither did other organizations, such as the World Bank and the Japanese government, apply any pressure. According to the World Bank's economist, Ghulam Qadir,[16] the predicament was also because Pakistan suffered from serious problems of 'governance'. He claimed that the situation pertaining to corruption in the country was very bad, comparing the situation with some of the poorer African countries. Responsible people such as parliamentarians and senior bureaucrats were involved in siphoning off resources allocated for development projects. It is true that for Pakistan the overall corruption was such a malaise it was questionable whether resources diverted from defense would be naturally and judiciously invested in social development. Reduction of defense spending, none the less, would give a boost to Pakistan's flagrant economy. A reduction would require serious planning for rationalizing military plans that were not on the cards during the period under study. Studies were being carried out within the military organization but without the necessary political will to do the same. During this period Islamabad did not show any signs of decreasing its military spending. The growth in the defense budget did slow down during the 1990s primarily owing to a decrease in major arms transfers, but nevertheless, there was resistance to a strategic reduction in military spending. As an essential part of the broader security or external threat debate, any discussion on the reduction of military expenditure acquires serious emotional and psychological dimensions. Successive governments, afraid of becoming unpopular, have not been able to attain any worthwhile reduction in security spending. Reducing military spending, in my view, would be tantamount to admitting defeat against India, which would have a drastic short-term effect for a ruling government. Focusing on such consequences, the political leadership or the policy-makers do not consider the fact that decreasing military expenditure and diverting it for socioeconomic development would have long-lasting implications for the nation's progress. Unfortunately, decision-makers have shirked from adopting a long-term approach that was also obvious from certain strategic decisions taken over the years.

5
Pakistan's Arms Suppliers

For technology-dependent nations military modernization poses a tremendous problem. It is a question of buying what is required and finding an appropriate source for military hardware shopping. Pakistan is no exception. Financial constraints put policy-makers at a loss to fulfil weapons modernization plans. The two options they always resorted to were finding a source that (a) would be willing to provide Islamabad with cost-free equipment, or at financially palatable terms, as part of some strategic alignment, and (b) would provide it with a credit facility. Pakistan's relations with China, US, and some European states were framed in the context of potential arms transfers.

Pakistan–US arms transfer links: 1979–88

Arms transfers between Pakistan and America formed the core of their relations.[1] Weapons supplied by the US, first during the 1950s and then in the 1980s, was the hallmark of Pakistan's military modernization. This was contrary to the American view of its South Asia ally which saw the Pakistan–US relations as essentially woven in the Cold War paradigm with Islamabad having a marginal utility for American policy-makers. By joining the US-sponsored military assistance pacts in the 1950s and 1960s, SEATO and CENTO, Islamabad had become part of a group of frontline states that were to deter the former USSR from any military adventures. This was a role that India has always declined to play. Countering the communist threat was precisely the factor that brought the US back to Pakistan in the 1980s.

The two allies had found a common ground for a temporary security alignment which ensured economic aid and military supplies to Pakistan in return for assistance Islamabad would provide in

Washington's covert operations against the Soviet troops in Afghanistan. Prior to 1979, Pakistan's significance for the US had diminished to that of a state with peripheral importance for American security interests, issues such as human rights abuse by Islamabad and its nuclear proliferation being the centerpiece of the diplomatic relations between the two countries. Under the leadership of Zulfiqar Ali Bhutto, foreign and defense policies had undergone a change. Bhutto introduced new trends in the country's foreign ties. His compulsion of projecting his figure as a Third World and or prominent muslim leader took him away from the US, a policy not pursued by the earlier government of Field Marshal Ayub Khan. American help was sought in the 1950s and 1960s not only to undertake the much-needed military modernization but also to acquire legitimacy for the military dictator's political venture. Bhutto had transformed diplomatic relations through befriending China and adopting a proactive stance in attracting some Islamic countries. Some of these countries were to provide funding for Islamabad's nuclear program. Unlike the pro-US leaders in Pakistan, Bhutto did not shirk from putting nuclear proliferation at the forefront of bilateral relations with Washington. This also put Islamabad on a collision course with Washington. The Carter administration had imposed an arms embargo (President Carter's presidential directive (PD-13)) that identified social problems as being more dangerous for international peace and security and was instrumental in blocking assistance to Pakistan.

1979 can be viewed as a watershed in Pakistan–US relations when Washington reversed its earlier policy. This put nuclear proliferation and other issues on the back burner. This policy change had occurred during the last days of the Carter administration and offered Pakistan $400 million to improve its defenses against any possibility of a Soviet attack. General Zia rejected the offer imputing it as 'peanuts'. The Pakistani military dictator had realized the importance of the strategic development represented by the Soviet invasion of Afghanistan and its impact on Cold War politics. He was sure that the Americans would 'come down on their knees'.[2] General Zia had a great ability to assess and play on the psychology of people including nationals and foreigners. Lt General Chisti, one of Zia's closest aides, believed that Zia had manipulated the situation emanating from the Afghan crisis to gain some strategic advantage and he further explained that Zia had postponed the signing of 1978 agreement with Afghanistan's President Daud who was to be removed soon after. What had instigated Zia's stance? Did he have some prior knowledge of the developments that were to take place and the impact they would have on US–Soviet relations as well as on American strategy in the South Asian region?

Zia was definitely an intelligent and perceptive person who knew how to cash in on an opportunity provided by Moscow's invasion in Afghanistan. He was certainly not privy to Soviet plans of invading Afghanistan, but was quick in perceiving the sensitivity of the moment. Particularly after the invasion took place, he decided to manipulate conditions in a way that derived maximum benefit for Islamabad. The Reagan administration played along in allowing Islamabad to convince American Congressmen that Pakistan's security was threatened.[3] American policy-makers and the defense bureaucracy were determined to punish Moscow for its transgression against the norm of East–West relations established after the Second World War.[4] The US aid offer to its ally was revised to $3.2 billion, a package that comprised economic and military assistance programs. The military component was worth $1.6 billion and used to procure hardware for the Pakistani armed forces. The most vital part of the first military assistance program was the 40 F-16s, USAF's top-of-the-line fighter aircraft. Under Secretary of State Buckley, testifying before the Senate foreign relations committee, explained the government's point of view. According to him: 'A strong, stable and independent Pakistan is an essential anchor of the entire West Asian region.'[5] Justifying the action of providing military assistance to a foreign government, he further added: 'The marginal US dollar loaned under FMS to the Thai or Turkish Army or Pakistan Air Force is a dollar that we otherwise would have to spend outright on our own forces to do a job that the Turks and Thais and Pakistanis can do better and at less cost.'[6] Such an argument helped in appeasing members of Congress who opposed the idea of transferring state-of-the-art technology such as the F-16s to Pakistan.[7] Their view was that the aid was a financial burden on the US economy, depleting some vital spares stocks and compromising the overall defense of America's West European allies. (On Pakistan's demand Washington had agreed on accelerated sales of the aircraft.)

The transfer of these aircraft not only indicated the success of Pakistani policy-makers' planning but also American desperation to turn Afghanistan into Moscow's Vietnam. Arms transfers to Pakistan not only reflected Reagan's ease with 'soldier turned politician,'[8] Zia-ul-Haq, but it also depicted a fear of communist designs that had to be averted through strengthening Pakistan. The US had invested approximately $2 billion and committed about a hundred CIA officers to aid the *mujahideen* fight the Soviet troops.[9] In addition, an array of weapons was provided to Islamabad (see Table 5.1) mainly to reward Pakistan's government for the help it rendered in carrying out insurgency operations in Afghanistan.

Reagan's approach was based on the objective of gaining a clear strategic advantage over the traditional foe. The presence of Soviet bases in South Yemen and Ethiopia, and loss of America's strong ally in the Persian Gulf littoral, the Shah of Iran, were perceived by American policy-makers as a reduction in US military capabilities[10] – a view shared by the American public. There was a growing paranoia within the US of increasing Soviet military superiority. This period, termed by Halliday as a 'second cold war', saw an increase in anti-Communist sentiments in the US, especially the Southern and Western regions which were centers of right-wing politics and America's military industrial complex.[11] According to Hathday the popular anti-Communist perceptions provided a specific direction to foreign policy at the time. Weapons transfer to designated allies was one of the prime instruments of American policy and it was at this juncture that F-15s were sold to Saudi Arabia.

Henceforth, nuclear proliferation, which was not entirely forgotten by American policy-makers, ceased to be the only issue at Capitol Hill in ties with Pakistan. Lt General (Retd.) Ijaz Azeem, posted as Pakistan's ambassador to the US in the early 1980s, said that Islamabad's uranium enrichment was given secondary importance, although some Congressmen were agitated about what Pakistan was doing on this score.[12] Reagan had hoped to 'kill two birds with one stone' through providing conventional arms to Islamabad. Given Pakistan's security concerns the supply of weapons would eliminate Pakistan's need for non-conventional defense, a policy obvious from James L. Buckley's testimony before Congress in 1981.

> We do believe that our best chance to influence the outcome, influence the future direction of what might be Pakistani intentions, is to help remove the very significant sense of insecurity that the nation suffers from today. We believe that if real insecurity can be removed we will not only have a better chance to make sure that explosives are not detonated, but also would be in the best position to use the argument of persuasion that this would not be in Pakistan's best interest.[13]

The American analysis was proved wrong in the ensuing years. Successive governments in Washington did not realize that deep-seated security concerns of a nation could not be assuaged through inconsistent policies and short-term assistance. Also, there was little appreciation of the internal dynamics that contributed towards nuclear proliferation. General Zia-ul-Haq, ever watchful of American policy-making, played along by giving the impression that the US approach towards containing nuclear proliferation in the region might work. During an interview on an American television network he said: 'Pakistan would not make a bomb if the US continues with its assistance to his country.'[14]

Islamabad used American anti-Communist sentiments to justify its requests for military hardware. According to Lt General Safdar it was quite easy to obtain weapons from the US during that period. It was all a matter of convincing the Americans that these were required against the USSR.[15] Not that everything Islamabad demanded was given, but certain major weapon systems that Pakistan had never hoped to get in the past were obtained. This situation continued until 1988 when a series of politico-strategic developments changed the *status quo*.

Pakistan–US arms transfer links: 1988–99

After 1988 American policy towards Pakistan started to drift back to the earlier position. The withdrawal of Soviet troops from Afghanistan, begun after the signing of the Geneva accord, followed by the collapse of the USSR putting an end to the Cold War, made Pakistan strategically less vital to the US. This caused a divergence of perceptions on key issues that led to a breakdown in the level of communication that had existed prior to 1988. It also resulted in the drying up of the American source of arms supply to Pakistan. The second aid package that was to commence that year for a six-year period was stopped because of the arms embargo that was imposed in 1990 subsequent to the introduction of the famous Pressler Amendment of the US Foreign Assistance Act. The new law required the American President's verification on Pakistan's non-involvement in nuclear proliferation. President Bush refused to provide such a guarantee. The end of the 1980s was the time when voices were being raised in the US regarding nuclear proliferation in South Asia and other parts of the world, and these became even more audible after the Soviet threat had completely receded. The US policy on Pakistan underwent a drastic change. Older and more controversial agendas were revived. Pakistani scholar Pervaiz Cheema termed this as the 'let-down' by the US.[16] This turn in American foreign policy was indeed disappointing for the Pakistani

Table 5.1 Pakistan–US Arms Transfers, 1979–98

Year of order	Year of delivery	Weapon designation	Type	Receiving service	Quantity
1980	1980	Gearing class	Destroyer	Navy	2
1981	1982–86	F-16A/B	Fighter aircraft	Air force	40
1981	1982–83	M48A5	Tank	Army	100
1981	1983–84	M109A2 155MM	Self-propelled gun	Army	64
1981	1983–86	BGM-71A Tow	Anti-tank missile	Army	1005
1981	1984–85	Model 209 AH-IS Cobra	Helicopters	Army	10
1981	1984–85	M110A2 203mm	Self-propelled gun	Army	40
1981	1984–85	M88A1	ARV	Army	35
1981	1984–85	M901 Tow	Tank destroyer	Army	24
1981	1984–86	M198 155mm	Towed gun	Army	75
1982	1982–83	Gearing class	Destroyer	Navy	2
1982	1984–85	M109A2 155mm	Self-propelled gun	Army	36
1982	1984–85	AN/TPQ-36	Tracking radar	Army	9
1982	1986	Model 209 AH-IS Cobra	Helicopter	Army	10
1984	1985	M48A5	Tanks	Army	35
1985	1985	FIM-92A Stinger	Portable SAM	Army	100
1985	1986	RGM-84A Harpoon	Ship-to-Ship missile	Navy	16
1985	1986	RGM-84A Launcher	Ship-to-Ship missile launcher	Navy	1
1985	1985–87	AIM-9M Sidewinder	Air-to-Air missile	Air force	500
1985	1985–89	Boeing 707	Transport aircraft	Air force	3
1985	1986–87	M113A2	APC	Army	110
1985	1986–89	M109A2 155mm	Self-propelled gun	Army	88
1985	1987–89	AN/TPQ-37	Tracking radar	Army	4

Table 5.1 Pakistan–US Arms Transfers, 1979–98 (*continued*)

Year of order	Year of delivery	Weapon designation	Type	Receiving service	Quantity
1986	1987	Model 209 UH-IB Huey	Helicopter	Army	3
1986	1987–90	BGM-71C 1 Tow	Anti-tank missile	Army	2030
1987	N/A	BGM-71D Tow 2	Anti-tank missile	Army	2386
1987	1987	FIM-92A Stinger	Portable SAM	Army	150
1987	1987–88	Phalanx	CIWS	Navy	6
1987	1987–88	RGM-84A Harpoon	Ship-to-Ship missile	Navy	20
1987	1998	P-3C Orion	ASW aircraft	Navy	3
1988	N/A	M109A2 155mm	Self-propelled gun	Army	20
1988	1988	Phalanx	CIWS	Navy	1
1988	1989	M198 155mm	Towed gun	Army	20
1988	1989	AN/TPQ-36	Tracking radar	Army	5
1988	1989	RIM67A Launcher	Ship-to-Air missile	Navy	4
1988	1989	RIM67A/SM1 standard	Ship-to-Air missile	Navy	64
1988	1989	UGM84A Harpoon	Surface-to-Ship missile	Navy	8
1988	1989	Brooke class	Frigate	Navy	4
1988	1989	Garcia class	Frigate	Navy	4
1988	1990	RGM84A Harpoon	Ship-to-Ship missile	Navy	18
1988	1990	RGM84A Launcher	Ship-to-Ship launcher	Navy	1
1989	1989	SH-2F/G Seasprite	Helicopter	Army	6
1989	1989	Phalanx	CIWS	Navy	4
1989	1989	Ajax class	Support ship	Navy	1
1989	1990–91	M113A2	APC	Army	775

Source: Stockholm International Peace Research Institute Database.

establishment. Although Islamabad was never unaware of the inconsistencies in Washington's nuclear proliferation policy and well understood the fragile nature of Pakistan–US security linkage, the withdrawal of Soviet troops and its impact on American policy-making was deemed sudden. Almost all segments in the Pakistani defense forces developed resentment against the US that was to increase during the 1990s resulting in a breakdown of communication.

The focus on containing nuclear proliferation that started from the Bush administration, however, was qualitatively different from the perception in the 1970s. It was underwritten by the new threat of 'rouge' states in maintaining peace in the world as well as questioning the US vital security interests. The role acquired by the West, particularly the US, of a 'benign hegemon'[17] did not allow other states to proliferate. The new western strategic perspective did not exhibit any consciousness of Pakistan's national security requirements or the fact that Islamabad had tried to draw Washington's attention towards the threat of nuclear proliferation in South Asia in the 1970s before India had carried out its first nuclear explosion in 1974. American nuclear nonproliferation policy was ridden with inconsistency. Islamabad managed to develop a nuclear weapons capability during the 1980s when Washington chose to look the other way and ignore Pakistan's activities. Hence, when Bush decided not to provide guarantees to Congress regarding Pakistan's nuclear agenda, his administration failed to dissuade Islamabad from proliferating. Abating Pakistan's security concerns was difficult without making India change its stance, which was much more popular with nuclear weapon states. This, at least, was a popular perception in Islamabad. It was felt that once Pakistan agreed to nuclear disarmament there would be no pressure on India to do the same. The pressure on Pakistan was comparatively more because of its relatively greater economic and military dependence on the West.

Realizing the limits of its non-proliferation policy, the Clinton administration modified its perception from calls to completely renounce non-conventional defense capabilities by the 'threshold' states to 'rolling back' or 'capping' the uranium enrichment program. The change occurred through a realization by Washington that nuclear threshold states could not be convinced of giving up their capabilities entirely. The US, therefore, would be content if these countries would stop further proliferation and halt uranium enrichment and weapons programs at their current levels which could be verified by the United States.[18] America's 'stick-and-carrot' policy *vis-à-vis* Pakistan had failed. Despite the arms embargo Islamabad was not willing to give up its nuclear option. The failure of coercive diplomacy was due to the lack of understanding of Islamabad's security concerns. The additional 72 F-16s for which Pakistan paid $658 million were not transferred. In 1993 Washington tried to entice Islamabad into containing its nuclear proliferation by offering to release the aircraft if Pakistan carried out a verifiable capping of the nuclear program. Such a strategy had been used earlier by the Carter administration when it had offered Islamabad a fleet of 110 A-7 aircraft at the end of the 1970s as bait to dissuade the latter from purchasing the reprocessing plant from the French. This offer was withdrawn when the elder Bhutto insisted upon acquiring the plutonium reprocessing plant. The basic idea behind the new diplomatic tactic of re-offering the fighter aircraft in the 1990s was explained by a state department official: 'The basic premise is that you have to offer something worthwhile for the Pakistanis to pursue this.'[19] This approach worked for a while when the nuclear program was capped as a result of the communication between Presidents Ishaq Khan and George Bush in 1990. This, as the ex-Army chief General Baig stated, did not include future R&D and the development of delivery systems. Pakistan was soon to abandon its compromising stance owing to lack of consensus in decision-making circles over the issue. The dividends of following American instructions were less than those of maintaining nuclear deterrence. The fact was, American policy on nuclear non-proliferation in South Asia put more pressure on Pakistan. Washington had absolutely no control over India's nuclear plans. New Delhi has always been averse to the idea of NPT rating it as biased against the threshold states.

The rewards offered to Pakistan were not substantial enough to persuade its policy-makers from not proliferating. The Brown amendment passed in 1995 permitted the transfer of certain armaments and spares to Pakistan excluding the F-16s. The aircraft were withheld in order not to disturb India. By the 1990s, the US State Department was inclined towards appeasing India. The new turn in US–India relations worried Islamabad. In the post-Cold War era, New Delhi emerged as a country of relatively greater significance for Washington and economic factors compelled the US to boost ties with New Delhi.[20] What worried Pakistani policy-makers was that India's significance had a security dimension as well. Washington had *de facto* recognized New Delhi's regional dominance – something that has always been unacceptable to Islamabad.[21] The signing of the deal between Rajiv Gandhi and Ronald Reagan on the transfer of dual-purpose technology, electronics and super-computers reinforced this particular anxiety. In addition, the US and Indian navies held a joint naval exercise in May 1992. Also, unlike the past, the US aircraft carrier USS *Missouri* was allowed to visit Bombay in 1987 which was evidence of the changing American view of South Asia and the level of reciprocity from India.

The new security equation worked out by American policy-makers 'added-on' India without subtracting Pakistan either.[22] The Brown amendment was introduced during this stride. Washington's approach towards South Asia adopted particularly in the 1990s was a more careful policy. Despite the recognition of Kashmir as a disputed territory and a crisis that needed to be resolved, in which the US would be willing to play an active role, Washington abstained from taking sides with any of the concerned parties. This policy appeared to be based on a realistic assessment of American influence in the region, and the desire to contain conflict without siding with either India or Pakistan. In the summer of 1999 Washington appealed to both countries to resist aggressive action during enhancement of tension.

It was obvious that Islamabad no longer enjoyed the status of a front line state. During the Gulf Crisis, America was careful in excessively encouraging Islamabad and giving an impression that Pakistan would be treated as a partner in the American security network. The Pakistan–US cooperation during the Gulf Crisis, when the Pakistani government was forthcoming in dispatching 5000 troops at Washington's request, was a temporary arrangement.[23] Joint commando exercises were held several times from 1990–97 but with limited objectives. The idea was to maintain a certain level of confidence in the Pakistani administration and its links with the US, but without creating any strategic liabilities for the latter. On the other hand, Islamabad's longing for a permanent military alignment runs very deep. Unlike India, Pakistan has always needed a foreign source that could be used as a cover to bolster its own credibility and strength while negotiating or dealing with New Delhi. Therefore, policy-makers in Islamabad liked to believe that Washington continued to be interested in Pakistan for strategic purposes after the end of the Cold War. There was also talk of American interest in turning Karachi into a separate state owing to what was believed to be American long-term commercial and strategic interests. This, it was felt, would sustain Pakistan's status as a frontline state for Washington.[24] The argument, though debatable, gives the observer a view of popular perception in Islamabad.

The two countries did not agree on the future of Afghanistan either. Prior to his death, General Zia had formulated plans of his own regarding the fate of the neighboring country. He had hoped to use the neighboring territory to provide strategic depth to Pakistan. He therefore sought the help of the *mujahideen* in a prospective conflict with India, and included it in an independent security network in which Pakistan would hold a prominent position. This he could do by neutralizing the government in Islamabad's favor. Resultantly, he was unhappy with the outcome of the Geneva accord that allowed the Soviet Union to pull out without establishing a pro-Pakistan regime. He, none the less, continued his plans that were persisted in by his successors in the Army. This was in accordance with the policy framework that was laid out in a paper produced by one of the branches of the Pakistan Army.[25] GHQ's plans supported the muslim fundamentalist groups in Afghanistan, a move not endorsed by Washington.

In 1996 the *Talibaan* were launched into Afghanistan. This was a group of fairly young religious fundamentalists who were trained in Pakistan and sent to Afghanistan to control a chaotic political situation. The Pakistan Army provided the backup. The *Talibaan* were successful in countering pro-Iranian and Russian groups in Afghanistan. These people were hard-core fundamentalists who tried to oppress all the freedom, liberalism and secularism that had existed in Afghanistan even during the days of the Soviet invasion. By 1997 Washington had grown critical of Islamabad's support of this group. This was not all – the group's existence threatened Pakistan society as well since the *Talibaan* influenced and intensified existing religious fundamentalist tendencies in Pakistan. This was one of the issues on which Islamabad and Washington's views clashed. Islamic fundamentalism tended to increase in Pakistan during this period, which had a definite anti-west and anti-US element to it. This added to American apprehension of the threat posed by Islamic fundamentalism.

Islamabad was quite conscious of this thinking. The inability, nonetheless, to control the military from pursuing an independent policy on Afghanistan led to the adoption of a two-pronged approach pursued by two different actors. On the one hand were the political governments that wanted to project Pakistan as a 'mildly' religious state. For instance, in October 1995 there was news of a failed fundamentalist coup staged by a handful of military officers. According to Islamabad, the situation was completely controlled, and the officers involved were arrested. The release of the news coincided with Prime Minister Benazir Bhutto's visit to the US and the debate in the American Congress on the Brown amendment. It benefited the government to project Pakistan as a moderately Islamic state where the democratic regime was entirely in control of running the country. It would be ambitious to suggest that such a strategy would pay great dividends to Islamabad or help restore Pakistan–US relations to the state prior to the enforcement of the Pressler amendment by President Bush. On the other hand, there was the Army supporting the religious extremists in Afghanistan.

A related issue was an increase in terrorism that seems to have been sponsored by the ISI. There were instances when Islamabad tended to cooperate with Washington, but this help however was limited to the catching of terrorists rather than containing them. In 1996 two American diplomats were shot dead in Karachi. The Pakistani establishment, especially the Army, was forthcoming in tracing the terrorists. Again, in 1997, the Army and the President were in the forefront of efforts to catch the terrorist Aamil Kansi who was involved in the murder of CIA officials in Virginia. He was extradited to the US. The operation for capturing the man was well planned and coordinated between American and Pakistani authorities. The fact that Islamabad did not follow the national extradition law in handing over the accused to the US created great furore in the country, but Pakistan's civil and military authorities ignored the public's hue and cry. It was through such acts that Islamabad hoped to maintain its existing links with Washington, if not get closer. The ultimate objective, nevertheless, was to build the confidence of American policy-makers sufficiently to re-open the doors of military assistance. This, at least, was the notion expressed by the Vice-Chief of the Air Staff.[26] His comment did not depict an appreciation of the fact that the fear of religious extremism would bring the US, Russia, India and the Central Asian Republics together. By 1999 a number of states were fearful of the *Taaliban* factor and the religious sentiments they espoused in the muslim populace of these states, resulting in internal security threats in different parts of the world. The philosophy was to establish Islamic rule through destroying the political systems in these countries. Even China, Pakistan's old ally, was concerned about the issue. It was no secret that the ISI, if not the political government, was supporting the fundamentalist faction in Afghanistan.

The divergence of ideas was also caused by the communication gap between the Pakistani establishment and the US that grew after 1990. Islamabad's nuclear option provided it with a confidence to pursue politico-strategic goals without accepting much foreign interference. It was believed that Washington could not afford to dictate terms to a nuclear Pakistan. This feeling was more pronounced among the military where there was an increase in anti-US sentiment.

Pakistan–Europe arms transfers links: 1979–99

With the drying up of the American source of weapons supply, Pakistan started to look to the West European arms market, where the natural choices were France and the UK with whom Islamabad had had past experience of arms transfers. These were the sources from which hardware could be purchased without any political conditions imposed by the suppliers. The acquisitions, however, were limited by the high price of the equipment (see Table 5.2).

Most of the equipment bought after 1988 was from France. Since the early 1970s, Paris has served as one of Pakistan's favorite options for arms acquisition, its relative desirability related to the absence of any political 'strings' attached to the sales. Despite the economic imperative of arms sales, Paris imposed a temporary embargo in July/August 1999 on the transfer of Agosta 90-B submarines. This sanction was imposed in the wake of fresh India–Pakistan hostilities in Kargil. The embargo was extended after the October coup in Pakistan.

Other major weapon systems were procured from the UK including frigates and some second-hand equipment, bought by the Pakistan Navy because its traditional inclination towards Western Europe, especially the UK. There were no political strings attached to these transfers either but problems were encountered later in keeping the equipment operational. Most of the British equipment used American components that were stopped after the arms embargo.

Another European source was Italy, albeit its offers were limited to sub-systems and components. It was primarily the two smaller services, the Air Force and Navy, which used this source. The sales were so insignificant that little notice was taken of these transfers. In 1996–97 the Army sought out Ukraine as a source of procurement of tanks and a deal was signed for the purchase of over three hundred. The East European source has much potential but the defense establishment and the government were unable to properly explore this market owing to an unprogressive foreign policy.

New sources

Although Islamabad's search for new supplies was limited, it managed to explore two other sources: South Africa and North Korea. It was mainly the Air Force and its organization, the Air Weapons Complex, that were involved in acquiring sub-systems from South Africa which

Table 5.2 Arms Acquisitions from Europe

Year of order	Year of delivery	Supplier	Weapon designation	Type	Receiving service	Quantity
1990	1995–96	France	Eridan class	Minehunters	Navy	3
1992	1995–96	UK	Type–21	Frigate	Navy	6
1994	1995–96	UK	Lynx	Helicopters	Navy	6
1994	1999	France	Agosta 90-B	Submarines	Navy	3
1996	1997–99	France	Mirage III	Fighter a/c	Air Force	40
1996–97	1998–99	Ukraine	T-80UD	Tanks	Army	320

were used primarily for the equipment produced indigenously by the Complex.

The North Korean source, on the other hand, was reportedly used to obtain missile technology for the indigenously produced ballistic missile, *Ghauri*. Not much information, however, was available on the procurement process.

Pakistan–China links: pre-1979

The main problem of acquiring armaments from a West European source was the relatively high cost. This factor alone was a profound problem for Islamabad which it tried to overcome by increasing the quantity of weapons. In order to change the focus from quality to quantity it turned towards its old ally, China.

Beijing had been forthcoming in providing hardware to Pakistan since the 1960s. In the past it was to reward Islamabad for aligning with China against New Delhi, it being vital for Beijing to militarily strengthen its South Asian ally. It was for this purpose that, in 1964, China had offered an interest-free loan to Pakistan worth $60 million. (This was after China's military conflict with India in 1962.) Another interest-free credit was offered in 1969 of $40.6 million, part of which was utilized to set up the heavy mechanical complex. Yet another loan of $217.4 million was extended during General Yahya Khan's visit to Beijing in November 1970. This aid was utilized for various development projects and helped Pakistan recover from a severe depletion of resources which had occurred as a result of the 1971 crisis in East Pakistan.[27] This placed Islamabad as the only non-communist Third World capital receiving huge assistance from Beijing. Noteworthy is that this aid to Pakistan was interest-free, and the weapons were free of cost.

Pakistan–China arms transfer links: 1979–99

The nature of bilateral ties underwent a change after 1979 when they became more focused on military technology transfers and arms production cooperation. This coincided with Chairman Mao's departure from the political scene. The end of the Cultural Revolution after his death heralded a new era in which economic rather than political considerations became more relevant to Chinese policy-makers. Also, this was a time that Beijing's relations with the US had started to become less hostile, as well as those with India. There was no dramatic turn in their bilateral relations, but Beijing was more willing to have friendly ties with New Delhi. During the 1980s relations improved to a degree that the two states decided to disengage some troops from their borders. This was done without shedding off key concerns about Tibet. The PLA general staff report continued to designate India as the most likely opponent.[28]

This change had a direct impact on Sino-Pakistan relations. After 1979 Beijing no longer considered it necessary to supply free weapons to its ally. Although these arms were relatively cheaper then the armaments acquired by Islamabad from the West, Pakistan was asked to pay for the hardware. Additionally, Islamabad had to pay interest for the credit provided by Beijing for arms purchase; Pakistan's procurement officials also found it increasingly difficult to negotiate the price of weapons systems with Chinese manufacturing firms. Beijing began to view its ally more as a potential source for earning foreign exchange. Pakistan's former Foreign Secretary, Sheharyar Khan, claimed that this was because a qualitative change had taken place in the bilateral links. 'The romance', he asserted, 'had gone out of the relationship'.[29] He further added that the Pakistan–China axis had become slack. China no longer vehemently supported its ally on the Kashmir issue or encouraged Pakistan in adopting a confrontational stance against India. Raju Thomas viewed this transformation in policy as Beijing's action of shedding its 'one friend–one enemy' approach for a 'two-friends' stance.[30]

Some Pakistani authors such as Rizvi held a different notion.[31] He viewed the period beyond 1979 as a continuation of the 1960s stance. In his opinion, any difference that came about was related to the change in Beijing's thinking on the resolution of the Kashmir issue, but other than that he saw no fluctuation. He based his argument on the exchange of high level delegations and similar issues. It is true that relations were cordial even after 1979 and they continued that way in the 1990s as well. China was supportive of Pakistan during the entire Afghan imbroglio with Deng Xiaoping and Zhao Zhiyang assuring Zia of their support when he visited China in 1982. Furthermore, hardware worth $3 794[32] million was given to Pakistan from 1979–92 (see Table 5.3). But these gestures merely portrayed the 'cosmetics' of the relationship and not the real policy.

Beijing continued to be the main source of nuclear material and missile technology acquisitions for Islamabad, but this must be viewed primarily in the light of military technology transfer relations developed during this period. Pakistan remained one of the biggest buyers of

Table 5.3   China–Pakistan Arms Transfers, 1979–92

Year of order	Year of delivery	Weapon designation	Type	Receiving service	Quantity
1980	1981	Hia Ying-2	Ship-to-Ship missile	Navy	8
1980	1981	Hia Ying-2	Launcher	Navy	4
1980	1981	Hegu class	Fast attack craft	Navy	4
1980	1981–82	T-60	Tank	Army	50
1981	1982–83	Type82 122mm	MRL	Army	50
1981	1983–84	A-5C Fantan	Fighter/ground attack aircraft	Air Force	52
1983	1984	Hai Ying-2	Ship-to-Ship missile	Navy	16
1983	1984	Hai Ying-2	Launcher	Navy	4
1983	1984	Huangfen class	Fast attack craft	Navy	4
1983	1986–87	F-7M Airguard	Fighter aircraft	Air Force	20
1984	NA	A-5A Fantan	Fighter/Ground attack aircraft	Air Force	98
1985	1985	HQ-2B	Surface-to-Air missile	Navy	20
1985	1985	HQ-2B	Surface-to-Air missile system	Navy	2
1985	1987	Fuqing Class	Support ship	Navy	1
1987	NA	K-8	Jet trainer	Air Force	6
1988	1989–90	Type P58A	Patrol craft	Navy	4
1988	1989–91	T-69	Tank	Army	275
1988	1989–92	Anza (under license)	Portable SAM	Army	350
1988	1990–91	F-7P Airguard	Fighter aircraft	Air Force	80
1988	1992	F-7M Airguard	Fighter aircraft	Air Force	40
1989	1990–92	Red Arrow-8	Anti-tank missile	Army	150
1989	1991–92	T-69II	Tank	Army	160
1990	1991	M-11	Ballistic missile	Army	55
1990	1991	M-11 Launcher	Ballistic missile launcher	Army	20
1992	NA	F-7P Skybolt	Fighter aircraft	Air Force	40

Source: Stockholm International Peace Research Institute Database.

Chinese technology: a factor that had overwhelming influences on policy-making in Beijing. Chinese military technology, it must be noted, was qualitatively poor with a limited number of buyers, Pakistan being one such recipient. These sales meant additional revenue for Beijing – a fact that could not be ignored. Thus, various technologies including the M-11 missiles were transferred to Islamabad. This was done despite Washington's annoyance and when the US government tried to turn this particular transfer into an issue, the transfer of Completely-Built-Units (CBU) was replaced by the transfer of the technology to assemble the missiles in Pakistan.

China's arms transfers during the 1990s reflected its desire to exert its policy of a regional power to deal with states, especially Asian countries like Pakistan and Iran. Providing sensitive technology to Islamabad, nevertheless, was based on slightly different calculations. Besides other reasons, one explanation was rewarding Pakistan for the help it rendered in supplying Western technology and aiding China in improving indigenous defense production. Beijing makes much of its array of weapons but its locally manufactured armaments are rather inferior. Better quality hardware was badly needed by the PLA, in accordance with the change of military strategy based on a 'forward projection',[33] but this could not be achieved by out-dated equipment. Pakistani military, particularly the PAF, had superior operational know-how to suggest modifications to the design of weapon systems. The two countries jointly undertook three major defense production projects during this period, two of which related to the Air Force. The real beneficiary of these deals was Beijing for there was no substantial transfer of technology to Islamabad that could help it improve its technological base. The transfer of equipment in semi-knockdown (SKD) form would earn foreign exchange for China without these transfers being recorded in the UN arms trade register.

Beijing's assistance was considered vital for developing Pakistan's poor defense industrial resources and technology. In the absence of a sound defense production capability, it was obvious that weapons transfers should become a significant issue in Pakistan's ties with China.

6
Military Industrial Complex

The literature on arms production identifies the existence of a hierarchical system based on three distinctive tiers of weapons manufacturers: first, second and third. Most of the less-industrialized nations, particularly the Third World, fall in the last category. A state's indigenous arms production capabilities determine its placing in the hierarchy of weapons-producing nations. This in turn ascertains the flexibility it enjoys in providing for its own defense.

Pakistan's defense production: an overview

Pakistani policy-makers' concern for territorial security and the urge to increase self-sufficiency in weapons was the prime reason for Islamabad's indigenous manufacturing. Like any country engaged in a protracted conflict, Pakistan's need for a sustained supply of conventional weapons was vital for its armed forces. Manufacturing weapons at home was considered a viable option to escape from the 'blackmail' of foreign arms suppliers. The objective was to attain self-sufficiency in weapons that would reduce national dependence on foreign sources. According to the former Army chief General Abdul Waheed, 'It is a time tested fact that no country can maintain her armed forces on borrowed weapons. To be self-confident in the community of nations Pakistan must become self-sufficient in defence production.'[1] A surge in indigenous production activities was noticeable in the 1970s and again in the 1980s. From 1979–99, the military industrial complex witnessed an expansion with a number of indigenous manufacture projects.

Rizvi cites the desire for the economic spin-off effects as a reason behind the creation of the military industrial complex[2] but, however, failed to present a strong case in support of his argument. There was no evidence of any broad-based economic or technological benefits of the military industrial complex. This was mainly because defense industrialization plans were not based on a strategy of industrial and technological growth with 'spin-offs' and 'spin-ins'. Indigenous production activities were instead established on a narrow-based approach of 'import substitution', which was a hallmark of Field Marshal Ayub's industrial development planning. The idea was to attain limited self-sufficiency in certain types of hardware to meet urgent needs and avoid a complete paralysis in a future military conflict. Islamabad's experience during the 1965 and 1971 wars was bitter. The American arms embargo imposed in the middle of the 1965 war had largely impaired the military's ability to fight more effectively than it had desired.

Since the 1960s, successive governments have officially endorsed this policy. In a presidential directive issued in 1985, President Zia ordered all arms acquisitions to be accompanied with the transfer of technology. What appeared to be a dynamic approach nevertheless did not boost defense production. The continued poor state of Pakistan's arms industry was linked with limited management objectives, mismanagement of defense industrial policy-making, and the state of the technological and scientific capabilities of the nation.

For two reasons self-sufficiency and import substitution proved to be self-defeating goals. First, defense industrial activities were limited to the assembly and manufacture of a few, technologically less advanced weapon systems. Islamabad could not markedly reduce its dependence on foreign sources for supply of major weapon systems especially when no efforts were made to expand the industrial base. The vendor industry, as will be evaluated later in this chapter, was not encouraged to help the public sector defense industry with assimilation of defense technology and expansion of indigenous manufacturing capabilities. Second, production activities were entirely focused on the needs of Pakistan's defense forces that meant limited production. This not only compromised industrial efficiency but also made production costly. Exports that could have allowed Pakistan to attain economies of scale did not play a major role in defense industrial planning with mainly only small arms and ammunition sold to approximately thirty countries including Sri Lanka, UAE, Bangladesh and others (mainly Middle East countries). The export figures were given around Rs. 400 million annually,[3] the only area with a certain level of excellence being the manufacture of small arms and ammunition. The Pakistan Ordnance Factories (POFs) is one of the two facilities where some reverse engineering is carried out and the Pakistan aeronautical complex is also involved in selling its services to friendly countries – the figures for which are unknown. Another facility to start limited commercial activities was the Air Weapons Complex (AWC), but its export activities have not expanded because of technological limitations. Nevertheless, commercialization of the defence industry was not a strong feature of Pakistan's arms manufacturing. The limited demand was also one of the reasons that policy-makers opted for plants of sub-optimal scale. In any case, almost all organizations operated on a one to one-and-a-half shift basis due to limited demand. This had an impact on the 'learning curve' that did not experience an upward slant. There were instances when a technology was obtained through a transfer arrangement but, because of huge gaps in production, the facility concerned eventually lost the expertise for sustained production.

The limitation of Islamabad's defense production goals was linked with the parochial nature of arms procurement policies. Governments were so focused on fulfilling immediate weapon requirements that indigenous manufacture was assigned secondary importance. Over the years the arms import lobby was also a major player hampering progress in the local defense industry. This lobby also made inroads into the defense industry and encouraged a culture of imports in the local weapons production facilities. A large portion of the resources allocated for indigenous production was spent on the import of raw materials and other necessary items. As a result a strategic partnership between the public sector and private entrepreneurs could not develop.

The limited market for locally manufactured hardware also affected the total cost of production. The formula of 'increased output reduces cost' was not adhered to, which was not surprising since the defense industry is 'budget' oriented and not 'profit' oriented. The federal government provided funding on an annual basis to the arms manufacturing concerns in accordance with the demands of the three services. The final products transfer to the armed forces without the involvement of any cash transactions. This makes it difficult for the defence industry to earn any profits. Also, arms manufacturing concerns use traditional methods of labor-cost calculation in which cost is equated with direct labor cost. The problem with this methodology is that it only takes into account the cost of production and ignores any non-producing costs. Furthermore, most of Pakistan's defence production concerns are over-staffed, thus making manufacturing even costlier. Although it is argued that, in the Third World countries, labor employed in the defence industries is cheap,[4] in Pakistan's context, however, over-employment was indicative of a negative employment-to-output ratio. This, therefore, liquidates the cost-effectiveness factor *vis-à-vis* the availability of a cheap workforce. Military managers of these industries had no notion of the impact of cost or economies of scale on production, one major drawback being the absence of a system for cost analysis. Under the circumstances, the top managers of the various facilities operate on the basis of a vague notion of taking advantage of cheap labor costs as compared with a number of industrialized countries.

Lack of quality control is yet another issue. Although ISO 9000 certification was obtained by the Mirage Rebuild Factory, the overall culture is not quality oriented. Such changes would require strategic reorganization that military decision-makers were not willing to consider, largely because of the incompetence of people controlling the production facilities and defense industrial planning who were ignorant of certain fundamental concepts in defense production and production management. Owing to the reactive and bureaucratic/organizational orientation of military officials, these various institutions have been transformed into 'hyper-bureaucracies' resulting in three obvious effects: (a) it led to the creation of large infrastructure and duplication of activities; (b) made production operations inefficient; and (c) hindered quality control mechanisms. According to the Army's former Chief of the General Staff (CGS), the Army was conscious that the performance of some facilities was not up to the mark,[5] although there is no evidence of such an argument followed up. No steps were ever taken to reduce the burden on the national exchequer by reorganizing defence production infrastructure, one glaring example being the establishment of the Naval Research and Development Authority (NRDA), which was established primarily to satisfy Prime Minister Sharif that the service was involved in R&D.

Like any third tier manufacturer Pakistan used the 'learning-by-doing' approach. This was considered cost-effective and an efficient way to progress in local weapons production. One Army general opined, 'Because we cannot afford to make everything ourselves, we have been forced to take a composite approach ... that applies both to our approach making an item of equipment and our need to persuade countries like China to cooperate in the development stage with us.[6] This methodology is based on the idea of 'on the-job' learning without investing large funds in R&D, a strategy adopted by countries where technology is not sufficiently developed and resources are scarce.[7] Hence, know-how is developed through importing and using technologies from industrialized states. Islamabad followed a similar path by initially acquiring the technology to overhaul and assemble particular weapons with the ultimate objective of beginning full-fledged manufacturing. The idea was to start with phasing out foreign components, which would help save foreign exchange, ultimately moving to a stage where most of the equipment could be manufactured locally. The second part of the plan could not materialize for almost total absence of R&D and poor planning. It is noteworthy that a certain level of R&D is necessary for the anchorage of any technology[8] and this was always a weak area in Pakistan's case because the government never invested sufficient funds. The three R&D institutions mentioned later in the chapter use most of their funds on personnel and maintenance, with little going into real R&D work.

The dearth of skilled manpower was another explanation for the scarcity of R&D. All the former and serving heads of defence production facilities and the R&D establishments, who were interviewed, complained about the lack of technically skilled people in the country. This situation represents the poor state of education, particularly the lack of technical education in the country. According to the World Development Report 1994,[9] only three per cent of the people in Pakistan were at the tertiary level of education in 1991. The condition of scientific education is even worse: according to the Pakistan Engineering Council Report for 1987–8 there were only 33 215 graduate engineers registered with the council including 2024 electronic, 7417 electrical, 7284 mechanical and 11 444 civil engineers.[10] However, there existed small 'pockets' of skilled people whose expertise remained under-utilized, for instance, skilled aeronautical engineers employed by the national carrier, PIA. In another case, Dr AbdulQadeer Khan claimed to have created a team of 1200 scientists and engineers at the Kahuta Research Laboratories. This team, it must be pointed out, was involved in the production of conventional shoulder-fired missiles since 1990–91. Dr Khan claimed that conventional weapons production had begun to make use of excess capacity and time.[11] It was, therefore, intriguing to hear people complaining about the lack of skilled manpower to carry out R&D. Indubitably, the under-utilization of this 'pool' of experts relates to bureaucratic and personal interests and errors in planning and control. A defense production board was constituted in the 1980s to plan and monitor all activities related to military technology transfers and indigenous manufacture, although the board obviously failed to carry out the assigned task.

Lax control indicated a lack of vision. A defense industry constitutes a niche in the overall industrialization of a state but this requires a metamorphosis of a socioeconomic system from agrarian to industrial. In Pakistan, where the decision-making elite is predominantly from the agricultural class, such a change naturally could not take place. Moreover, owing to the collusion of interests between military and civilian policy-makers, the armed forces were not motivated to bring about a change. Any alteration in the socioeconomic environment was difficult because of continued political instability in the country. The domestic political conditions were also discouraging for the suppliers of technical know-how. From 1979 to 1999 political changes were so rapid that any foreign supplier was at a loss regarding any fixed national policy on indigenous production or industrialization, let alone weapon manufacture. Resultantly, all areas of industrial development were neglected; foremost was the capability to produce strategic raw materials that were seriously lacking. For example, there is no facility in the country for the manufactures of steel alloy, the Pakistan Steel Mills and the Heavy Mechanical Complex not having the capability of producing steel alloys for the defence industry.[12] The Heavy Mechanical Complex at Taxila had obtained the technology to make gun barrels for tanks and to carry out other high-tech mechanical work, but much of its capacity remains unutilised. (The Complex is also dependent upon foreign sources of supply.)

The country also lagged behind in the important technology of metallurgy, the government failing to exploit the capacities of existing institutions such as the Metal Industries Research and Development Corporation. This was established for the benefit of both public and private sector industries but neither sector benefited from it. According to its Director, despite the capability of the research organization to make almost an entire tank with boron steel, and having the machinery to do work in the field of metallurgy, there has not been enough encouragement from the government.[13] Consequently, Islamabad imported the necessary basic raw materials increasing the overall cost of manufacturing.

This particular inertia was linked with the lack of an industrial base and culture in the country. Manufacturing has never been a strong national activity. There were eight to ten manufacturing houses in the entire country who were in the business because of family traditions. Bhutto's nationalization of industrial units in the 1970s had completely crushed private entrepreneurs. What further hindered industrial growth were serious infrastructure problems like shortage of electricity and bureaucratic red tape, making Pakistan's defence-industrial capacity shallow and non-dynamic.[14]

Shortage of resources was another failing.[15] With the military top brass's bias for foreign equipment and its short-term approach to acquire weapons from wherever possible and in the shortest time possible, there were no signs of any inclination to invest funds in indigenous manufacture. Islamabad tried to project a positive message for the private industry in the early 1990s, during Prime Minister Nawaz Sharif's first government, by offering huge loans for setting up defense manufacturing related production facilities. Moreover, he offered to permit private industry to share the use of public sector defense manufacturing facilities, but Sharif's public statements, nevertheless, failed to make a positive contribution to boosting confidence. The private sector's continued skepticism of the government was on three accounts: first, Pakistan's economic conditions were in a poor state and industrialists were unsure about Islamabad's ability to provide the resources that could help the private sector set up the infrastructure required for defense manufacturing.[16] The mixed signals sent by the conflicting statements of the Prime Minister and the Finance Minister further confused private entrepreneurs. Second, no statements from the government materialized. For instance, Islamabad had offered to restructure the duties imposed on the import of the raw material used by private industry for defense production but these were never revised. Third, industrialists feared that public sector inefficiency common in these facilities would prove infectious.[17] The public sector defense industry, especially the military bureaucratic culture prevalent in the facilities, is different from private sector culture, the military managers' need for control and discipline not meshing with the ways of the private entrepreneurs. Military bureaucrats, it must be remembered, are extremely skeptical of civilian involvement at any place where defense work is carried out. It is felt that these officials would constantly interfere with the work schedule.

What increased the financial problems was the general mismanagement of resources, also common in the defense industry. Although defense production had less budgetary allocation than arms purchases or other activities, the expenditure was not efficient because of lack of proper financial control. Funds were provided without any due care for evaluating the long-term benefits of an allocation. There were instances when machinery and equipment was imported from abroad without any proper planning for its utilization. Materials were procured on a 'as needed' basis without an efficient inventory control often resulting in redundant stores. The presence of corruption, at the same time, exacerbated the problems.[18]

There was no evidence of Islamabad's honest attempt to exploit sources other than government revenue to eradicate the problem of financing indigenous production. A possible source under consideration is the private sector that, in a number of cases, was effective in helping with R&D activities and weapons production projects. In 1991, in public–private sector cooperation in the US, 56 per cent of the $152 billion R&D expenditure was financed by the private sector,[19] the involvement of the private sector helping to reduce the burden of financing the activity from the shoulders of the government. Contrastingly, the Pakistan defence industry's interaction with the private sector is limited to insignificant activities. Although over a thousand subcontractors work in military production, this figure constitutes less than 20 per cent of companies whose business activities are more than 50 per cent defense oriented. The public sector industry is the prime contractor with little room for vendors to grow. The defense industry suffered from a lack of a multiple layered subcontracting base. The vendor industry is what formed the subcontractor's level, but they are not involved in the manufacture of 'high-tech' goods. These subcontractors can be classified into two groups: (a) the semi-government establishments who, mainly because of their connection with the government, are given the task of producing some components or building complete weapon systems. They mainly cater to specific and random needs of arms manufacturing establishments. The Machine Tool Factory, Precision Engineering Complex, and Kahuta Research Laboratories are involved in carrying out work for the POFs and the Army respectively; and (b) the private contractors, who are basically limited to the task of manufacturing small items.

The rules and regulations pertaining to the entry into this area of activity are stringent and discouraged private entrepreneurs, a fact that applies to defence production all over the world. In Pakistan's case the problem is acute mainly because bureaucratic red tape complicates procedures. The registration process, contract bidding, getting a deal, and the related payments are so long and tedious (the entire process takes more than two years) that discouraged companies left the business at the first available chance. In the defense-vendor relationship, there is a complete absence of a concept of a long-term relationship and vendors bitterly complain about the lack of support from government.

Military managers find it equally difficult to work with the private sector. They are of the view that private entrepreneurs are not quality conscious and have no experience of manufacturing high-tech items.[20] They also believe that the private industry shirks from committing funds of their own for R&D and manufacturing projects, claiming that they tend to depend on funds provided by the government. Considering the major concern of corporate sectors all over the world is profit-maximization, it would be unfair to expect this kind of patriotism from Pakistani industrialists. Regarding the issue of quality it is a problem linked with the general state of quality consciousness and control in the country. The major production activity of these defence subcontractors is for the domestic civil market and, like other manufacturers, they are used to producing substandard goods. The lack of quality persists because of the non-existent pressure from the market to improve the standard of production or services. This situation can only be improved through a long-term relationship in which public sector managers would ensure that work is carried out according to laid down specifications and quality standards. Generating detailed specifications is a task that defense managers did not fancy.

Private entrepreneurs, on the other hand, did not agree with the official contention, being of the view that the government's lack of understanding of their problems, together with bureaucratic red tape are highly discouraging factors.[21] In the case of shipbuilding and the ship-repair industry, the large number of agencies involved at Karachi port had scared away potential business. There is no forum for voicing these concerns and eradicating the problems. Private entrepreneurs were particularly shy of investing in R&D because they were not protected by patent laws, neither are designs or plans for modification in a project any guarantee of production orders and interestingly the cost of R&D is not reimbursed. The fact is that getting production orders really depended upon the personality of a military industrial manager and his inclination towards a particular company. If a particular official's policy is to encourage a certain vendor, then the entire process of selection and getting payments for the work would become less cumbersome. The absence of 'continuity' in the military bureaucracy is found to be equally damaging because the policy of the manager would not necessarily be pursued by his successor, a behavior that means uncertainty and confusion. The limited demand of the military is also a discouraging factor. What makes things worse is that when procuring from local sources military managers expected competitive rates. Private entrepreneurs are expected to acquire technology, invest in R&D and production, and yet to be able to sell at cheaper rates than foreign sources. The military management exhibits an inability to understand the fundamental dynamics of production, economies of scale, and its linkage with price. Neither do they realize that the tax structure does not provide protection for private industrialists to import raw material in order to manufacture defence-related items.

Defence Production for the Army

Pakistan's indigenous weapons production activities consist of a variety of activities and projects with most of the major projects started in the 1960s. New ventures were also initiated from 1979 to 1999, the majority of which were for/of the Army. The discussion in this chapter, however, does not differentiate between the old and new. In the following sub-sections all activities mainly dedicated to the Army will be analysed.

Pakistan Ordnance Factories (POFs)

The POFs was established in the 1960s with Chinese help to manufacture small arms and ammunition for the military. Situated at Wah, it comprises fourteen factories producing projectiles, infantry equipment and ammunition, explosives and even military clothing. The factories produce mainly for the three services of the armed forces. Limited production for exports and commercial sale was also carried out.

Inefficient production planning and management meant that the facility had turned into a 'white elephant'.[22] There were three basic explanations for poor performance: first, the factories were overstaffed, the number of employees being about 50 000, Islamabad's practice being to use its public sector to boost employment. Nepotism also played a major role in excess employment, a common feature of all major government departments. This policy contributed to the inefficiency of the factories. Over-employment increased the non-productive labor hours, thus adding considerably to the total cost of production. Secondly, the facility was never able to utilize its full capacity. For instance, the ammunition production units that operated at a capacity worth $30–40 million had a capacity worth $70 million. All the fourteen units of the POFs were operating on a one to one-and-a half shift basis. These low production figures are understandable as the facility produced mainly for the national armed forces. Manufacture for export does not alter the situation either. Claims were made to the effect that in the early 1990s the POFs exported goods worth $50 million.[23] The facility was given a semi-independent status with a relatively free hand in initiating projects and procuring raw materials and machines. This independence was not used properly by the managers, who indulged in wastage of resources earned from exports and domestic sales. The facility produced ammunition for .12 bore shotguns for sale in the domestic market but this has not proved profitable because of the poor quality. Small arms and ammunition dealers told the author that such ammunition produced by manufacturers from the private sector performed better than the POFs,[24] let alone later was more expensive. In Pakistan, where automatic and semi-automatic small arms flooded the market after 1980–81, the .12 bore gun did not remain a fashionable commodity.

The official position on the efficiency of the facility was different. For instance, the balance sheet of POFs for the financial year 1991–92 showed a 'break-even' position between revenue and expenditure. (The budget allotment was for Rs. 3.084 million against an expenditure of Rs. 3.082 million.) Such figures could hardly be considered reliable especially the way this facility was managed. As mentioned earlier, technologies, spares and materials were often procured without proper planning: a case in point relates to the decision to acquire technology to manufacture the 7.62 mm and 5.56 mm ammunition from the US. The original idea was to switch over from the production of 5.56 mm to 7.62 mm ammunition, which is standard for NATO. This plan was never put into action and the machinery imported for the purpose was left lying idle.[25] The Auditor-General's report published at the end of 1980s also pointed out certain inefficiencies. According to this report the sten gun ammunition production had been doubled in 1987–88, and 14.4 per cent of this ammunition had remained unsold in 1988–89 alone.[26] The piled-up inventory was the 'sunk' cost that added to the cost of operations. There was a definite problem with the choice of the inappropriate technology, a feature common to decision-making in most Third World countries.[27] The quality control system of 'random testing' used in the facility also reflected upon the management style. Despite a thousand inspectors deputed by the Army GHQ for the purpose, they were not able to ensure any improvement in quality.

The management was unable to introduce any significant R&D in the organization, although some reverse engineering and minor modifications were carried out during the 1980s, termed as R&D. For instance, the factories managed to copy the Russian 100 mm and 75 mm HEAT anti-tank ammunition, and RPG-7. A similar project was also carried out on the American 73 mm fin-stabilized rocket that was obtained during the Afghan war. Minor modifications comprised alterations in the butt of the German G-3 rifle where the original design was replaced with a retractable butt stock. In addition, the British made 105 mm TK Hesh explosive was copied to make the 100 mm explosive.

The lack of R&D was linked with shortage and mismanagement of resources, and the absence of a research culture. It is worth mentioning that the major chunk of financial resources for meeting the local military orders is obtained from the MoD. In addition, POFs retained funds generated from commercial sales, but these resources are not utilized prudently owing to poor management and decision-making, an example of which was given earlier. Neither was there any sign of the management trying to reduce costs by sharing work with private industry. The organization had 153 vendors who catered to a small portion of the demand for low-tech components such as fuses.

Heavy Industries – Taxila (HIT)

This facility was established in 1979, again with Chinese help, close to the POFs. The plan conceived in 1971 aimed at developing the capability to overhaul/rebuild Chinese T-59 tanks. The idea was to gradually learn how to produce an indigenous tank. The facility consists of five independent units: two of these were to overhaul the Chinese T-Series and American M-series tanks; one unit each was dedicated to the production and assembly of the main battle tank and armored personnel carrier; and the last unit was to manufacture the gun barrel.

The HIT also adopted a composite approach in assimilating foreign technology although R&D was not part of the core activities. Since work started in 1979, approximately 1000 T-59 tanks and tank engines were overhauled. These tanks were also up-gunned. Technologies for carrying out 'heavy duty' mechanical engineering work, such as shell casting, investment casting, gas nitriding, and a tool tip plant were obtained from China, enabling HIT to manufacture 9000 out of the 11 000 components needed for a T-59.

Subsequently, in the 1980s, a project to manufacture a main battle tank was launched. The idea was to build an indigenous tank for which the country could draw on its earlier experience of overhauling and rebuilding Chinese tanks. The plan was considered as an efficient way of obtaining a main battle tank for the Army being less expensive than the American M1A1 tank, and easier to manufacture because of the earlier relative experience. The design was derived from the marriage of existing blueprints of the Chinese T-85s and T-69IIs. This indigenous tank, known as MBT-2000 or 'A1-Khalid,' was to have a 125 mm gun with APFSDS, HEAT and HE ammunition. Additional features included improved armor, an upgraded engine, a laser range finder and a computerized fire control system. Instead of starting from the drawing board, as India had done for its indigenous tank, 'Arjun', this strategy would cut the cost. The adversary's experience had shown to Islamabad that a new design would prove expensive in terms of time, money and efficiency. The MBT-2000, therefore, was planned to use 45 per cent components from the previous models and the project was undertaken with Chinese support. A development contract was signed in 1988–89 with the Chinese manufacturer 'Norinco' that would spend a major portion of the total cost of the project. The estimated figure was approximately Rs. 25 billion ($1.2 billion) out of which Pakistan agreed to commit $1 billion.[28] It was only $435 million that Islamabad would spend on the design and development stage. According to the plans, the tank-manufacturing factory was made to cater to produce 150–200 tanks annually.

Claims were made to the effect that, 'Pakistan has attained the capacity of designing, developing and manufacturing tanks ...'[29] This was arguable for several reasons: first, the production part of the entire program was behind schedule. In September 1999 there was a spurious report of the tank being ready for full-scale factory production. Second, Pakistan had not yet mastered the art of tank manufacture. Major General Utra, who had served as the Director-General of HIT during the 1980s, argued that most of the work done at the facility pertained to mechanical engineering. The facility had a small electronics shop for minor repair work only.[30] Vital components were imported from foreign sources which lead to delay and quality electronics was not the strength of China's defense industry. Moreover, this also increased the total cost of production. In any case, the MBT-2000 related work at HIT would be limited to assembly work. The main battle tank was one of the many examples in which dependence on the OEM would continue. There were no signs of minimizing this dependence mainly because of lack of strategic industrial and technological capacities. The lack of synergy between collaborators hampered development and the perpetual dependence on the OEM would make self-sufficiency a self-defeating formula.

The third project, launched 1987–91 with the help of General Dynamics, was to overhaul and rebuild M-48A5 American tanks. The tanks were acquired from the US as part of its military assistance program ensuring commitment of $30–40 million.[31] People were dispatched from HIT for training at the M-48 rebuild factory in Arafiya, Turkey. A new factory was also constructed for this purpose, on the pattern of the T-series tank factory. The project was stalled by the American arms embargo. The termination of the project did not surprise certain people involved with HIT at different times. They challenged the logic of establishing such a unit in the first place. According to Maj. Generals (Retd.) Utra and Afzal, both of whom had headed the facility, the Army's workshop 502 at Rawalpindi was sufficient to carry out repairs and basic overhaul of American tanks. The Army's middle management was interested in retaining the smaller facility but the Ministry of Defence pushed for a new establishment which it could directly control and supervise. In the end it was the MoD that prevailed upon the Army's decision and moved the project to its present sight at Taxila. The decision was taken by General Zia-ul-Haq in 1985, justifying it on technical grounds. Zia hoped that with a full-fledged facility, the Army's plan to change the hull of the tank could materialize, something that could not be done at the 502 workshop.

Another project housed in the same premises was the assembly of armored personnel carriers. A contract was signed with an American company FMC in 1989 to assemble approximately 775 completely-knock-down (CKD) kits of American M113A2 APCs. Although this program was not entirely affected by the arms embargo, the pace of work slowed down not only because of the unavailability of all the kits, but this was a deliberate action as well. According to Maj. General Ahmed Ali, Additional Secretary (Defence Production Division, 1994), this was to keep employees busy for a longer period because the APC project was the only work being done in the factory. It was feared that quick completion of the project would result in inactivity[32] and these permanent employees could not be fired.

HIT suffers from a similar over-employment problem as the POFs. The total number of workers is approximately five to six thousand. Lt General Naqvi, Director-General HIT until 1995, claimed that the labor-to-output ratio was 10:1 by 1994. All three factories housed single plants of sub-optimal scale. This, the General said, suited Pakistan's requirements. As far as quality assurance at the unit is concerned, the facility has its own system for assuring quality with its independent team of inspectors.[33] Like the POFs, the facility has about 200 vendors, but only six or seven of them are actively involved. Business relations with the vendor industry are marked by cyclical changes – a relatively good period followed by a lull in activities. The times when buyer–supplier relations improved were the years that HIT top management was progressive.

Institute of Optronics (IOP)

The institute was established in 1985 at Rawalpindi to assemble/manufacture the night vision devices for the Army use. It was involved in the assembly of four types of devices:

(a) AN/PVS-4A (weapon sight)
(b) AN/TVS-5A (weapon sight)
(c) AN/PVS-5A (high performance goggles)
(d) GP/NVB-4A and GP/NVB-5A (night vision binoculars)

The main idea was to start from assembly to gradually progressing towards complete manufacture of the product at home. The planners had prepared a feasibility report in which it was suggested to the government that this facility would save almost 30 per cent of the foreign exchange needed for the procurement of the completely-built-units. The deletion and assembly process was started in 1988 and the first delivery to the Army was made in the same year.

There were flaws in the projection of 'deletion' attained by the facility mainly because 'deletion' was calculated on the basis of number of components rather than total price factor. Moreover, 'deletion' had only been attained in components that required mechanical engineering processes. No efforts were ever made or succeeded in manufacturing parts that required electronic and optro-electronic engineering. In fact merely the testing of imported tubes and lenses was carried out. The apparent inability of the institute is due to the lack of R&D. This in turn is linked with the shortage of resources and people with technical know-how.

The organization's allocation for FY 1993–94 was Rs. 10–12 million, almost half being spent on pay and allowances and an equal amount on general maintenance. The amount did not vary in the years before or after 1994. There were about 180–200 employees out of which only 35 worked for the R&D section.[34] Furthermore, these employees included only one PhD – with this kind of budget and manpower it was not possible to carry out any profound research work.

Margalla Electronics (ME)

Based in Islamabad, the establishment was created in 1985 to manufacture ground radar. The birth of ME was more a result of bureaucratic/organizational tensions and manipulations than strategic needs. As a result, in nine to ten years only 40 radar were assembled. There was no real indigenous production done there. The facility was to attain a 'deletion' target of 70 per cent by 1994. Though a whole section was created for the purpose, in reality this target was never achieved. The basic work carried out was reverse-engineering the cards (computer/electronic chips) of the various systems. Considering the slow pace of work, the government had thought of closing down the facility in 1991, but the idea never materialized owing to the interest of a few who immediately imported a few CKDs for assembly, and the facility, therefore, was never wound up.

The Yasoob truck project

After 1985 Islamabad approved the project for the local manufacture of military trucks. The Army in particular was interested in two specifications: 4 × 4 and 6 × 6. The program was not undertaken at any existing public sector production facility because there was none to carry out such a task. The governments' effort to develop the automobile industry in the 1960s and 1970s had failed and by the 1980s there was no automobile engineering base for the military to bank on. Help was sought from the private sector and a collaborative project of the private and public sectors was undertaken. An organization namely Trans-Mobile Ltd (TML) was established with the government being represented by a semi-governmental corporation, PACO. The first prototypes for the 6×6 and 4 × 4 were approved by the Army in 1991 and 1994 respectively and the Army, being one of the end-users, was involved with the testing and inspection of the vehicles. The service had placed an order for approximately 3000 trucks deliverable in five to six years. According to the defense production division, TML planned to make one truck per day.[35] Major components such as engines, gearboxes and so on, were to be imported from abroad, the local industry having absolutely no capability to make certain components. Therefore, most indigenous input comprised mechanical engineering and fabrication. TML's top management claimed that there were 'buy-back' arrangements with the foreign companies regarding certain components. There was an overall plan to start 'deletion' of foreign components and the programme was divided into stages, 40 per cent to be attained in the first stage. *Jane's* also confirmed this figure.[36]

Like the arrangement between TML and the government, the company was to be allowed to export as well. The company's top management hoped that, despite the profit margins not being very high, they could make the project fairly profitable. Nevertheless, bureaucratic red tape became a major hindrance in the company's plans and the firm was not able to increase production and utilize its production capacity to the optimum. TML's production line continued to operate on a single shift basis.[37] Consequently, there were delays in production with military officials blaming the private sector for this cunctation. It was as late as 1997 that Yasoob trucks could be spotted in Army units.

Missiles and military electronics

During the 1980s several programs were launched for the manufacture of laser range finders and missiles such as the Chinese Red Arrow-8 and Swedish RBS-70. In addition, work was also begun on an indigenous missile 'Anza-II.'

The manufacture of the laser range finder was ascribed to Dr AbdulQadeer Khan and his organization, the Kahuta Research Laboratories. Reports on the LRF 786P laser range finder described it as a medium range, hand held, lightweight and rugged device. It is of compact size, simple to operate, and versatile for ranging both static and moving targets. Its range was given as 150 m to 15 km; however, there were no reports on its performance. Sources from the armed forces claimed that the equipment was made with imported components, an opinion also expressed about the indigenous missile 'Anza-II.' This shoulder fired anti-aircraft missile has the same specifications as the American Stinger missile, has heart-seeking capacity and was also made under Dr Khan's supervision. Although Islamabad claimed that the missile was co-developed with Chinese help, dissenting sources claimed Pakistan made no major contribution in the development of 'Anza-II.' Dr Khan's setup had claimed the complete manufacture of the medium range ballistic missile, Ghauri, as well.

These missiles, supposedly produced by Pakistan, were actually assembly work with 70–80 per cent components imported from Beijing or North Korea.[38] During this period Islamabad also produced the Red-Arrow-8 anti-tank missile, the technology of which came from China as well. Again, Dr Khan supervised this licensed production. Another missile, the Swedish RBS-70, was manufactured at the Precision Engineering Complex, Karachi. This missile was procured from Sweden in 1986 in CBU form, and later an agreement was signed for the transfer of technology. The Precision Engineering Complex was given the task because of its capability to conduct relatively high-tech work. The organization worked as a sub-contractor for the Ministry of Defence. The Swedish missile was basically being assembled from imported components. The mechanical engineering work, which required precision engineering, was done at the facility in Karachi. This assembly reportedly began in 1988–89 and approximately 1000 missiles were manufactured up until 1997–98.

Defence production for the Air Force

The Pakistan Aeronautical Complex (PAC) Kamra is one of the two manufacturing facilities of the Air Force. The four independent units of the facility are dedicated to the overhaul and rebuild of Chinese and French aircraft in the PAF inventory, as well as to the manufacture of light aircraft and assembly of ground-based radar.

The idea originated in the early 1970s for the overhaul of the Chinese F-6 and French Mirage aircraft. The facility started its operations in 1978, when the first Mirage was overhauled. The technology acquired from the French was housed in one of the first major units called the Mirage rebuild factory which employs approximately 2000 engineers and technicians who can completely overhaul the Mirage aircraft and its Atar 9C engines. The facility's current overhaul capacity (of Mirages) stood at 8–10 aircraft and over 50 engines annually. The officials at Kamra said that the Mirages were overhauled after 600 flying hours but this was increased to 800 hours by carrying out certain modifications. By 1997 the factory overhauled an aircraft in nine months or 248 days in eight different stages. It is argued that the time taken for the overhaul of the Mirage, if the aircraft were to be sent to France, would be eighteen to twenty months.[39] The facility also overhauled the Mirage IIIO aircraft procured from Australia. These were second hand aircraft, and the Air Force had only bought these for cannibalization. However, owing to the efforts of the aeronautical complex, a large number was retrieved for use.

The officials of the facility also claimed that PAC was involved in direct offsets. Approximately 139 parts for the Mirage were being made for supply to France under a 'buy-back' agreement. Some parts made for the 'Mushshak' were also sold to the Swedish manufacturer. Neither of the two arrangements was impressive, especially the one with the Swedish manufacturer. Saab had sold a few of these aircraft to Pakistan and to a private company in Switzerland with the OEM hardly carrying out any production in the 1990s. The factory also tried to sell its services to the UAE by offering the overhaul of Mirage aircraft. The UAE government was asked to procure the necessary spares from the French that were not received in time. As a result the exercise was not repeated. The Director General, PAC, claimed that the French had delayed the supply on purpose to discourage the UAE or any state from going to other sources for the Mirage overhauling.[40] This case depicts the dilemma of a typically dependent third-tier manufacturer. In the absence of the capability to manufacture vital spares, Pakistan's hands are tied in extending such a service to friendly states. Furthermore, work was also delayed because of the sub-optimal size of the overhauling unit. In 1996–97 the PAC had to contract the French company, Sagem, to overhaul and upgrade around forty Mirage IIIs. This was for two reasons: first, these aircraft needed a level of upgrading that the aeronautical complex could not do; second, the Mirage rebuild factory was already busy overhauling about thirty-six Mirage aircraft. The factory, nevertheless, was the first to introduce the quality control tool of ISO-9000.

In 1988 the technical capability for upgrading and overhauling the Prat & Whitney F100-PW-220 engines of the F16s was acquired from the US. The plant obtained for approximately $40 million was placed in the Mirage rebuild factory. The capability was not part of the US FMS program, hence was part of a commercial deal signed with the OEM. With this capability the F-16 engines could be upgraded from an 1850-hour engine cycle to 4000-hour. The Director-General PAC in 1994 asserted that work was formally started in 1991. He also claimed that PAC had offset arrangement with General Dynamics and worked as one of the approved vendors for the manufacturer of the F-16 but this was not confirmed by any American source.

Another important facility at Kamra is for the overhaul and rebuild of the Chinese F-6 aircraft. The idea was launched in 1972 and it was formally established in 1980, primarily to overhaul the aircraft, with Beijing providing the technological assistance. In the 1990s the facility was in a position to manufacture 7000 spares for the F-6, FT-5, and FT-6 aircraft. In addition to the spares it manufactured different varieties of fuel drop tanks for these aircraft. The factory's capability to overhaul the different types of Chinese aircraft was enhanced from eight to 24 warplanes per year. Each aircraft is overhauled in 30 to 45 days. Moreover, the facility also overhauled the F-7 aircraft.

The third unit, the aircraft manufacturing factory, was rated as one of the most significant projects at the facility. The factory has the capability to manufacture the propeller driven Swedish Saab MFI-17 aircraft and its derivatives. The Saab MFI-17 was re-named as 'Mushshak' and its improved version Shahbaz. By establishing this facility the PAF high command had hoped to enter indigenous defense production whereby they would learn to manufacture a complete aircraft. The eventual plan was to make sophisticated jet-engine aircraft. There were about a thousand people working at the facility when it started operations in 1981. The factory has the capacity to manufacture 24 such aircraft,[41] although actual production rates can vary. A total of 243 'Mushshaks' were assembled up to 1994.[42]

As for the grand strategy of aircraft production, the plans did not materialize, the factory only producing 'Mushshaks'. The aircraft's technology was radically basic. The PAC staff took pride in the development and production of 'Shahbaz' that is only an improved version of 'Mushshak'. The main difference between the two is that 'Shahbaz' uses a more powerful American engine. The aircraft had some minor shortcomings in its design, for example, the air-conditioning unit which is fitted behind the back of the pilot's neck causing inconvenience during flight. Furthermore, limited production caused the cost of production to be relatively high. The fly-away-factory cost for 'Mushshak' was listed as $185 000 and $200 000 for 'Shahbaz'. The facility managed to export some of these aircraft to friendly countries, a considerable quantity of which being provided free of cost. The prospects for developing a local market for such aircraft are also bleak. Private entrepreneurs claimed that for the price of this local aircraft they could buy two foreign-built aeroplanes with better technology. One reason for price escalation was the dependence on foreign sources for the kit of materials, especially major components like the stuffing, gear-box, engine, instruments, and so on. The 8000 parts produced at Kamra were mainly comprised of sheet metal and some machine parts.

Other projects undertaken by the facility with Chinese assistance, such as the co-development and co-production of a jet trainer, Karakoram-8 or K-8 and Super-7, met a similar fate. The Chinese aircraft manufacturer CATIC conceived the original idea for the jet trainer which, at that time, was known as L-8. Later, when Pakistan was included in the project, the name was changed to K-8. Some of the Pakistani officials serving at PAC in 1994 claimed at the time that the idea had originated in Pakistan and the initial design work was carried out at the aeronautical complex.[43] This does not seem likely because the PAC does not have any R&D or design capability, although it is possible that Beijing consulted the Pakistan Air Force on the design of the aircraft. Pakistan's 25 per cent share in the aircraft was limited to the airframe for which an investment of $6 million was made in 1986. There was a total lack of offsets and the Chinese manufacturer CATIC was the prime contractor. In 1994 plant was set up to cater for the production of 467 minor parts for K-8. The possibility of PAC expanding the production target was not encouraging since the Chinese were not forthcoming in transferring the technology related to the manufacture of electronic or more sophisticated parts. The engine selected for the aircraft was the American Garrett TFE731–2A–2A. There was a certain tension between the Chinese and Pakistani authorities over the choice of engines – the PLA Air Force wanted its own engine, while the PAF opted for the American Garrett engine for its better performance. The PAF also wanted to install other quality systems like Martin Baker MK-IOL rocket assisted ejection seats and Collins EFIS system for the cockpit. The disagreement on the choice of these systems was one reason that the design could not be frozen and by 1999, the project was still in the development stage. With the current specifications it is hoped that a jet trainer could be produced at a cost of $2.1 million.[44]

One of the four factories is the Kamra Avionics and Radar Factory (KARF), established in 1987 to assemble radar and other electronic equipment. The factory assembles pulse-doppler radar and associated power generators under license from the German company, Siemens. The Managing Director of the factory agreed that it was only the L-band low-level ground-based radar and power generators that were to be assembled at the factory.[45] The technology of this radar is not very sophisticated. In any case, most of the activity at this unit was assembly work. Approximately five radar are made every year for which all the electronic components are imported. Activities at KARF are, therefore, a clear duplication of work carried out at Margalla Electronics. Air Vice-Marshal Yusuf Khan's remarks to justify the duplication were that KARF was an effort to build an empire.[46] The few vendors involved with activities at the PAC were doing low-tech work restricted to low shelf-life components such as parts of the batteries used for operating the radar.

Air Weapons Complex (AWC)

Created around 1993–94, the Air Weapons Complex denotes a slight change from the traditional public-sector facilities in the country and ensures profit maximization by adopting a commercial approach. The organization is controlled by the National Weapons Complex, a supreme regulatory authority also in charge of the National Development Complex (NDC). Headed by a serving Air Force officer and manned by civilian specialists, the AWC is supervised by the same board of governors as the NDC. The Prime Minister chairs the board of governors. The basic formula was to anchor core technologies and develop indigenous production by making the exercise cost effective. This objective was achieved through developing dual-use technologies. AWC sold its products to other public sector concerns such as the Pakistan Telecommunication Corporation Limited. For this, the organization was given financial autonomy. The Air Headquarters provides basic control mainly in the form of financial support and limited manpower. For operations, civilian experts are hired at higher rates than are offered to public sector employees. The establishment also has an independent marketing wing to sell its products locally and internationally. Furthermore, it has the autonomy to procure the necessary systems and components from any source found feasible. One of the pioneering projects launched in 1986–87 was to reverse-engineer the American Sidewinder missile and manufacture an indigenous IR-guided Air-to-Air missile.

Defence production for the Navy

The Naval Dockyard situated at Karachi is the main facility for naval weapon production. In 1947 it was a small facility with a workforce of 500 people to carry out repair work on naval vessels. This was later expanded in 1952 to undertake rebuild and repair work. In 1997 the number of employees was 8000 (excluding 83 engineers); it has 66 workshops, four floating docks and one graving dock. Although the work carried out at the facility is only for the Navy, it was brought under the direct control of the MoD in 1993–94. This development was not explained. The authorities claim that the Naval Dockyard has the capacity to simultaneously build two major ships, three missile boats, and two submarines. The Naval Dockyard also produces gun- and missile-boats of Chinese origin. According to official reports, the activities are divided into two broad categories: (a) the overhaul and rebuild of different vessels of the Pakistan Navy, and (b) indigenous manufacture of certain systems. Senior officials of the facility said that up to 1994 37 major ships and 14 submarines were rebuilt, and nine different types of naval vessels were overhauled.[47] Manufacturing activities at the dockyard were limited to small projects and consisted of the construction of over 45 small vessels and some floating docks which have the capability of handling vessels up to 1000 tons.

In the early 1990s the Navy signed two separate deals with the French regarding the assembly of mine hunters and submarines at Karachi. In both cases the manufacturing work was to be limited to assembly with components imported from France. These projects did not enhance technical know-how, hence production work at the Dockyard was limited to manufacturing floating docks and overhauling a limited number of midgets. Another project, launched in 1997, was for assembly of gun- and missile-boats, which were to be made under a transfer of technology agreement with China.

Karachi Shipyard and Engineering Works (KSEW)

Established in the 1960s to boost the naval manufacturing industry, KSEW was originally controlled by the Ministry of Production. It was brought under the Navy's control in the 1980s to ensure proper co-ordination between the Navy and the shipyard. KSEW was then constructing hulls for French submarines and missile-boats. Consequently, naval officers took over the management, but the labor force remained predominantly civilian. There were subsequent plans to move all naval construction activities to KSEW, bearing in mind the export potential. The most significant case was of the Chinese F-22 frigates for which negotiations were carried out with a Chinese shipyard to transfer the technology for the construction of these frigates to the KSEW. An upgrading of the shipyard was also part of the negotiations.

One explanation for bringing the facility under military control was to ensure improved efficiency. Naval officers being good disciplinarians, it was believed, would manage to tame the rather unruly civilian labor force. The KSEW's malady, however, was more complicated than simple labor problems. The lack of business, absence of a production master plan, bureaucratic red tape, the inability to market products and services, and the absence of an environment congenial for ship repair and construction were some of the many reasons that the facility could not perform. In 1998 the shipyard was termed as a financial defaulter with liability of Rs. one billion. Owing to bureaucratic control of the organization, the management had consumed over Rs. 400 million worth of workers' allowances. According to estimates presented by Citibank, an investment of Rs. 700 million was needed to relieve the facility from its status as a defaulter. According to government rules, a financial defaulter was barred from imports even though these might improve financial conditions. Under these circumstances it would not be possible for military managers, who had no notion of production and marketing, to lift performance off the ground. The performance had diminished from 72 orders in the 1960s (including four foreign orders), 47 in the 1970s (15 foreign orders), 39 in the 1980s (10 foreign orders), to 18 in the 1990s (seven foreign orders). The details of these orders were even more unimpressive. The average tonnage of the orders completed in the 1990s was about fifteen tons. The biggest order was 23 641 GR tons – this was a Chinese order that was stalled because of mismanagement. Despite an independent design section there was no original designing work carried out, most of the production being based on designs bought from abroad. In fact, a German manufacturer was one of the causes of financial liabilities by providing designs for the sea vessels.

Research and Development establishments

There are four main R&D establishments:

(i)   Defense Science and Technology Establishment (DESTO)
(ii) Military Vehicles Research and Development Establishment (MVRDE)
(iii) Armament Research and Development Establishment (ARDE)
(iv) Naval Research and Development Authority (NRDA).

DESTO was established in 1963 for research and development work related to improving weapon systems and conducting tests and trials. A civilian scientist headed the organization and managed a staff of a 1000 people comprising 220 scientists and 280–350 para-technical staff. In 1994 the establishment had an annual budget of Rs. 100 million of which 33 per cent was spent on pay and allowances.[48] A major portion of the remaining amount was spent on maintenance, barely leaving anything for the real activity of R&D. Under these circumstances, the main technical activity done by DESTO was limited to reverse-engineering and certain work in the field of chemical weapons.

The MVRDE was found to be no different. It grew out of a project launched in 1972 for the development of an indigenous tank. Originally called the Fighting Vehicle Research and Development Establishment, or Project-711, it was renamed the Military Vehicles Research and Development Establishment in 1974. Its officially proclaimed objectives were to carry out R&D for upgrading old military vehicles and designing new ones but this was a task not carried out at the facility. The basic work being done pertained to seeking out manufacturers from the private sector on behalf of the HIT. Small projects, mainly initiated by the HIT, are awarded to private industry through the MVRDE. Indubitably, this work can be done by the Heavy Industries itself. Other work that is done at the research facility was the construction of bridges and the like.[49] The establishment did not carry out any activity regarding design and development of military vehicles. Given the organization's financial and human resource constraints, it would be difficult to carry out real R&D. The facility has approximately 313 people and a budget of Rs. 10–12 million annually, out of which a large part was spent on pay and allowances.

The ARDE was created in 1974 to assist the POFs with research and development activities. Its main objective was to carry out major armament modifications and study the small arms industrial potential in the country. So far ARDE's activities have been limited to interacting with the private industry and providing them with the specifications of the demands of the POFs. The lack of any serious R&D is reflective of the general environment in the country that is not encouraging for R&D work. The organization's activities were limited to some reverse engineering[50] and again, like the MVRDE, the establishment is a duplication of certain activities that are being carried out independently at the POFs. Moreover, any form of R&D would be difficult with limited financial resources. It receives an annual allocation of around Rs. 16–18 million, out of which a large portion is spent in maintaining the staff of approximately 137 people.

The NRDA was established in 1998 in response to Prime Minister Nawaz Sharif's concern for establishing a naval R&D organization. Some of the young officers working for the Naval Dockyard were posted to the NRDA but with no clear objectives. The organization was involved in holding the first naval defense show in early 1999.

It is quite obvious from the description of the defense industry that the state of indigenous weapons manufacture is not encouraging. This primarily is linked with the absence of R&D and an industrial culture. The dearth of skilled manpower, and technological and industrial capabilities added to the problem. Under these circumstances, Pakistan will continue depending upon foreign sources for weapons supply and this situation is likely to persist unless or until Islamabad commits more resources to R&D, human resource development and industrialization as well as seriously improving its defense industrial planning.

Part II

In this part, consisting of four chapters, numerous arms procurement decisions carried out from 1979–99 have been analysed. These twenty-one years signify two periods: years when equipment was procured and technologies developed as a result of American assistance; and years when Pakistan depended upon its own resources. Pakistan's economic realities, foreign alignments and organizational interests of the defense establishment underwrote these decisions which were the result of the interplay between the various factors that were discussed in Part I. This part also contains an assessment of the future direction of military posture and relevant buildup.

7
Military Buildup Decisions, 1979–90

During this period Pakistan's military buildup was dependent upon American source of supply. The US military assistance begun in 1982 improved Pakistan's technological position against India with India seeing its strategic calculations being threatened by both the adversary's arms acquisition and the development of a non-conventional defense capability. The arms race that ensued between the two neighboring states also led to an increase in tension, which persisted well into the 1990s. This was the overbearing politico-strategic environment in which decisions were made.

Pakistan's military buildup, 1979–90: an overview

During this period Islamabad concentrated on strengthening its conventional weapons capability as well as building a nuclear capability. This was required in order to maintain a certain strategic balance with India that could enable Pakistan to defend itself. Given India's size and military technological advantage it was not possible to gain a decisive victory but the objective was to provide safeguards against a humiliating defeat, as was the case in the 1965 war. Conventional weapons procurement in these twelve years represented three distinctive periods: (a) 1979–82, when Islamabad was dependent upon its own resources to procure weapons that did not allow substantial arms acquisitions; (b) 1982–88, when Pakistan found the American doors reopened allowing it to acquire state-of-the-art military technology. Although the US military assistance was not sufficient to fulfil most of the needs of all the services, it contributed tremendously towards enhancing Islamabad's military capabilities. For the first time in Pakistan's military history state-of-the-art equipment was being obtained from a western source, making this transfer a vital technological injection in the South Asian region; and (c) 1988–90, when arms transfers from the US were cut off by the arms embargo imposed on Islamabad by Washington.

The period from 1979–90 was fairly barren with hardly any procurements carried out. The earlier arms embargo imposed on Pakistan for the 1971 conflict with India was relaxed in 1976, but Islamabad had to pay for arms imports from the United States and there were insufficient resources available to acquire hardware from the US, a source considered unreliable by the Bhutto regime. A couple of old Gearing class frigates were acquired on lease in 1980 but this was an insignificant transfer. In addition, Zulfiqar Ali Bhutto's nuclear proliferation policy had put him at cross-purposes with the US, which had further diminished chances of arms transfers. President Carter stopped all assistance to Pakistan, which led to Bhutto's famous remark about Pakistanis preferring to eat grass than giving up the nuclear option.

The change of heart in Washington was caused by the strategic development on Pakistan's northern borders. The American door reopened mainly to strengthen Pakistan against Soviet forces present near Pakistan's northern borders. Islamabad made sure that it contributed in deepening American fears of Moscow's aggressive policy and expansionist designs, a strategy that enabled the Pakistan government to procure hardware to strengthen its defense capabilities. There was an absolute clarity in the minds of the policy-makers that this injection of weapons would primarily cater to the Indian threat. This plan of action worked until the departure of Soviet troops from Afghanistan and the collapse of the USSR.

The start of American aid to Pakistan was also the beginning of a strategic thinking based on two extremes of the strategic spectrum. On the one hand was low-intensity conflict that was incorporated in military operational planning. The proxy war fought in Afghanistan had left deep impressions on the minds of planners who would try this technique on other fronts as well. On the other hand was nuclear conflict for which nuclear capability was developed. This capability was essential to maintain a strategic balance that was otherwise hard to keep in the region. American military assistance was never considered sufficient to tip the military balance in Pakistan's favor but the projection of the threat from the former USSR was useful in obtaining American support. Hence, most of Pakistan's arms procurement decisions during this period were necessarily linked with the US strategic perception of American security interests in South Asia.

The set of decisions made for utilizing this aid package focused on adopting an approach of narrowing the overall military capabilities gap *vis-à-vis* India, but this also meant that resources could not be equally divided between the three services of the military, an issue which further increased the inter-services rivalry and which became a more pronounced element after 1988, when the procurement discrepancy increased with the drying up of the American source. In the last three years leading up to 1990 very few acquisitions were made. This was due to three reasons: first, during the seven years from 1982 to 1988 Pakistan had linked its procurement planning with the American source and it would take time before weapons acquisition planning could be reformatted to bring in diversification. Second, policy-makers continued to hope to find a way of obtaining hardware that was stuck in the pipeline because of the embargo. Had the US released weapons, Islamabad would not have had to look at other sources, at least, for some years. Third, the country's economic situation had started to deteriorate rapidly.

Arms procurement for the Air Force

During the 1980s, the main focus of decision-makers was to strengthen the Air Force in the belief that enhancement of air power could offset the military's disadvantageous position in other areas. With army generals involved in domestic politics, it was considered strategically advantageous to strengthen the PAF instead. One of the lessons learnt from the two earlier wars with India was that a capable air force could prove effective in providing close-battle-support and necessary firepower to the ground forces. It could also enhance the military's capability to launch an offensive inside enemy territory. The largest portion of the American military assistance program, especially the first one, hence was spent on the Air Force. Of the military component of $1.6 billion, $1.2 billion was spent on the acquisition of 40 F-16 aircraft. These were transferred under two contracts: Peacegate I and II, and financed under the FMF scheme. The aircraft were delivered to Pakistan within 24 months by diverting them from the assembly line set up for Belgium and Holland. Again, at the time of the negotiations for the second aid package, a substantial amount was dedicated to the procurement of an additional 72 such aircraft.

The aircraft represented quality technology that the PAF had sought for years in order to narrow the technological gap with India. Prior to 1982, the PAF's mainstay were the Chinese, French and American aircraft of old vintage. The PAF had considered buying the French Mirage-2000 but the price was prohibitive. American aircraft were equally out of reach for political reasons. The frantic search came to a halt when the Reagan administration was willing to sell F-16s. These aircraft were preferred to any other aircraft for technological, financial and political reasons. Technologically, they improved the PAF's qualitative edge over its adversary in an unprecedented manner. The aircraft with its deep penetration, high manoeuvrability and multi-role capabilities enabled Pakistan, for the first time, to 'look the enemy into its eyes'.[1] Increased use of composite material in the airframe construction, fly-by-wire controls, high/g tolerance cockpit and a high-visibility bubble canopy made it an attractive option. The aircraft's armament and avionics such as the AN/ALQ-131 ECM pods, radar warning receiver, the ALR-69, the Thomson-CSF Atlas II laser designator pods, and the heat-seeking AIM-9L Sidewinder missiles were of special interest to the PAF. These technologies gave F-16s real clout and explains Islamabad's insistence on signing a conditional security deal with Washington tied with the provision of at least two of the special avionics systems.[2] After some resistance, the F-16s were finally transferred to Islamabad. Initially, it was the A-7s that were offered by Washington, but rejected by Pakistan on the grounds that they could not handle the enhanced threat. Eventually, pressed by its desire to seek Pakistan's help against Communist forces in Afghanistan, the US agreed and all systems other then the Thomson-CSF Atlas II laser designator pods were fitted into the aircraft and transferred to Pakistan. The laser designator pods were purchased later in 1983, in a commercial deal of $35 million, and fitted into 8–12 aircraft.[3]

The PAF did not take much time to assure the Americans that their policy had been correct. These aircraft were quite effective in taking care of Pakistan's air space violations by Soviet/Afghan aircraft. The 'rules of engagement' did not allow 'hot pursuit' of PAF pilots into Afghan territory and stipulated that the attack on the enemy aircraft was to be carried out in such a manner that the wreckage must fall inside Pakistan. The superior avionics, Beyond-Visual-Range (BVR) capability, and 'Sidewinder' missiles made this possible. From May 1986 to November 1988, the PAF succeeded in bringing down seven to eight Afghan aircraft. These results could not have been achieved by Chinese or French aircraft. Observing such an impressive performance, Washington rushed about 100 'Sidewinders' to Islamabad in 1985 – not a special consignment, but part of the 500 missiles that the US had approved for sale to its ally for $8.5 million. The Pakistani Air Force, impressed by the performance of these missiles, launched a challenge in 1986–87 to the POFs to reverse-engineer and manufacture the missiles – a high priority project undertaken by AWC staff. There was news of completion of the first flight-test in 1989,[4] but there was no news of the project's completion. This project was later formalized in the early 1990s.

The transfer of F-16s upset the Indians. In Mrs Gandhi's views, 'Pakistan had no "legitimate" defensive needs for such a sophisticated aircraft ... and when you make such a tremendous jump [in weapons technology] from one era to another you obviously make problems for your neighbors'.[5] The Indian Prime Minister put the blame on Pakistan and the US for initiating an arms race in the region. She obviously did not take into account India's military arsenal. Her main concern at the time was to check Pakistan from creating ripples in New Delhi's security planning for the South Asian region, which in her calculation should effectively be dominated by India. The F-16s increased Pakistan's capability and demonstrated Islamabad's will to counter India's military moves. General Zia's assurance to Mrs Gandhi that a limited number of the F-16s cannot make Pakistan any stronger could not abate her concerns.[6] Soon after the Pakistani acquisition India responded by acquiring MiG-29s from the Russians in 1983–84.

The F-16s proved vital in improving the morale of the Pakistan Air Force, in fact, of the entire nation. The aircraft was popularly viewed as a technology guaranteeing the territorial integrity of the country. Indubitably, the aircraft had become the center of gravity for the Air Force and gave it the clout it wanted. It was in order to maintain this edge that a decision was taken to enhance the F-16 inventory to 110 aircraft. An order was placed in 1986–87 for another 72 aircraft, starting the delivery from 1992. Of these, 11 aircraft were to replace the aircraft lost in attrition, for which Islamabad agreed to pay the commercial rate. Another 61 were to be paid from the second aid package for which Pakistan accepted a loan on concessional rates of interest. The total value of these aircraft was $1.4 billion. The second aid package was $4.2 billion of which the military component was $1.74 billion. It is obvious that a major part was to be spent on the Air Force.

This transfer was to be carried out under two schemes: Peacegate III and IV. Peacegate III comprised the transfer of eleven aircraft that were financed through FMF while the 61 aircraft that were part of Peacegate IV were to be paid from FMS. To accommodate the purchase of these aircraft, an approved deal for 75 Chinese F-7s in 1989 worth $225 million was cancelled.

Up to 1993, Islamabad paid a total of $658 million to the Americans. The payment should have been stopped in 1990 when the arms embargo was re-imposed, but in fact it did not, in the hope that the aircraft would finally be transferred to the PAF. The service played a major role in the continuation of this payment. Islamabad demanded return of this payment from the US – a contentious issue resolved in 1998 when Washington agreed to return some of the amount in the form of wheat.

The PAF top management had set its eyes on the F-16s as a replacement for the 170 ageing F-6s due for retirement. It was imperative for the PAF to replace these Chinese aircraft: a need clearly endorsed by the US administration as long as Soviet troops continued to be in Afghanistan. The Pakistani Chairman Joint Chiefs of Staff Committee's claim that Islamabad would not have acquired any other aircraft had it obtained the additional F-16s represented the Pakistan government's satisfaction with its decision to adopt these aircraft as the quality component in its Air Force's inventory.[7] What one fails to understand is that, by the 1980s Pakistan had sufficient experience of American-imposed embargoes to have gone for the same source to carry out procurement of a major weapon system on which the PAF based its entire operational plans. The military top brass was so preoccupied with addressing the technological gap with India that no thought was given to the possibility of an embargo. Military planners in Pakistan have always shied away from consulting the Foreign Office or individual experts for obtaining input on the non-financial aspects of the viability of procurement. What was more intriguing in the F-16s case was Islamabad's decision to pay the commercial rate of interest for the first aid package. Agha Shahi, then the Foreign Minister, claimed that this was to maintain Pakistan's independent foreign policy stance.[8] This stance was not adopted for the second aid package.

The F-16 deal had a prominent political dimension to it. The fact that Washington had agreed to supply its top-of-the-line aircraft to its South Asian ally was construed as a symbol of American support to different regimes in Pakistan – both the military and civil governments. It was of special value to General Zia's military regime. Zia required legitimacy for his government, which in the absence of support from within the country, needed endorsement from outside. The military assistance from the US, particularly the aircraft, provided him with popularity, especially among his main constituency – the military. It must be remembered that the Carter administration's offer to Pakistan of 110 A-7 aircraft during Bhutto's last days, which was later withdrawn, did not go down too well with the armed forces. The withdrawal of the offer symbolized ceasing the US support to Bhutto's government.[9] This was an important factor in the Pakistani military's 'change-of-heart' towards Zulfiqar Ali Bhutto's regime. The military generals attached great importance to the transfer of these aircraft because the US also gave these to Israel – a nation that has always enjoyed favorable treatment from Washington. Later governments recognized this significance too. The Federal Finance Minister of the Nawaz Sharif regime in 1997 held that the government realized that the F-16s held back by the US were an old technology that would not ameliorate PAF's strategic position, but that Islamabad was interested in these aircraft primarily because of the political imperative. The transfer of these aircraft would be a litmus test of Pakistan–US bilateral links.

Another sensitive technology that Islamabad tried to acquire during the 1980s was the Airborne Early Warning (AEW) aircraft. Initially, the E-3C AWACS were requested in order to improve the PAF's capacity to monitor and counter air space violations by Afghan/Soviet aircraft. The air space violations were threatening because they demanded retaliation. By implementing American security agenda in the region, Zia was inviting the wrath of Soviet hardliners, thereby taking a great risk. General Zia's strategy was partially successful in that the Reagan administration considered transferring AWACS to South Asia. The basic debate at that time was whether to sell the aircraft or to lease them to Pakistan. The aircraft cost more than Pakistan could afford either through the US military assistance package or through its own resources. Washington even toyed with the idea of temporarily transferring its own AWACS stationed in France or Saudi Arabia to Pakistan,[10] but this idea was abandoned for fear of downgrading American military preparedness, and giving Islamabad a technology that was too sophisticated for the South Asian region. Before completely dropping the idea of providing the AEW technology to its ally, other options, such as leasing the less sophisticated version of the Lockheed E-2C Hawkeye, were considered,[11] but this idea did not materialize either. After the Soviet withdrawal from Afghanistan, procuring this technology was unimaginable. Its procurement would have worried the Indians, who felt that this technology would have been directed against them. Even when negotiations were going on, New Delhi felt that the prospective Pakistani acquisition of these aircraft was to be directed against them. Indian defense analysts were of the view that since the technology would not prove effective in the mountainous terrain of Pakistan's northern borders, for which it was being demanded, it was bound to be used to keep track of India's forward deployment.[12] The PAF's procurement of these aircraft would certainly have been a greater technological leap than the F-16s.

During this period the PAF tried to strengthen the quantitative element as well. To cater to its procurement objective of a 'low-and-high-mix', two types of Chinese aircraft were procured, the A-5s and F-7s. About 90 A-5s were obtained in 1983–84 for a price of $1 million per aircraft which was a good price for a ground attack aircraft of this quality. The transfer of overhaul and rebuild technology was additional bait. (The first overhaul of these aircraft was reported in 1988.)[13] Another deal signed with Beijing during the 1980s was for the Chinese MiG-21, also known as the F-7. Since the 1970s the PAF had been interested in obtaining an efficient point defense/interceptor aircraft, a role that the F-7's limited 40-minute-endurance capacity could perform sufficiently.[14] Approximately 95 F-7 series aircraft were acquired in two orders: the first for 20 F-7Ps carrying 24 technical modifications to meet PAF's specifications and the second for 75 F-7MPs to carry more modifications. The Air Force wanted the Chinese manufacturer to supply the F-7MPs that had ground attack capability, but provided with the F-7Ps and F-7Ms instead.[15] The Air Force was not exactly satisfied with the Chinese manufacturer, Chengdu's F-7 also known as the supersonic sports plane. This caused them to look at the relatively cheaper option of upgrading the aircraft's technological capabilities through adding on components and technologies. Although officials, who were working at Air Headquarters, were critical of the policy of procuring cheaper aircraft and then spending additional resources on upgrading, this was the most logical path that could be followed by a resource constrained country.

In order to implement this approach the American manufacturer, Grumman was commissioned in 1986–87 to study modifications to F-7s that were necessary to eradicate the technical shortcomings of the Chinese aircraft. It is noteworthy that, because of certain serious technical deficiencies, the aircraft was initially not very popular even with the PLA Air Force because the Chinese-built systems suffer from crude manufacturing tolerances making the systems prone to failure. PAF officials, conscious of this problem and anxious to eradicate it, initially provided about $2 million to Grumman to study the possibilities of certain modifications. The PAF wanted the aircraft to be fitted with a General Electric F-404/100 or a Turbo Union RB 199 engine; a Wastinghouse APG-66 radar in a solid nose cone with lateral engine intakes in the wing roots; enlargement of wings to be fitted with leading-edge slots and extra 'Sidewinder' launch rails; replacement of the original ejection seat with a Martin-Baker ejection seat; fitting a F-20 style canopy and F-16 type HUD; and a 23 mm cannon. This project, designated as the 'Super Sabre' and later the 'Sabre II', failed and was abandoned. Air Chief Marshal (Retd.) Jamal Khan, while giving the reasons for the failure of the project, asserted, 'We did not get the plane that we wanted for the price we had visualized.'[16] The estimates given by the aircraft manufacturer were more than the $7–8 million per unit that Islamabad had bargained for. The escalation of estimated cost was due to technical reasons: one was that the manufacturers found it difficult to marry the engine with the airframe within the price range that Islamabad could afford. There was a serious problem of non-standardization of the airframe that made it difficult to apply a systems integration approach to all airframes, which were different from each other. There were other reasons as well. The timing of the study coincided with growing Sino-American tension after the Tiananmen Square massacre, and the Pentagon was averse to the idea of facilitating such technological help to the Chinese. It was due to a combination of these reasons that the project failed and the total amount of the study cost to Pakistan was around $3–4 million.[17] It was later that the PAF carried out gradual improvements in the F-7s by installing certain Western sub-systems.

The quantitative strength was increased, however, by procuring 50 Mirage IIIOs from Australia in 1990. These aircraft were retired from the RAAF, and most of them had about a hundred flying hours remaining on their airframe. Some of these were cannibalized in order to make a total of thirty aircraft operational. While representing a 'poor man's option', this was an excellent deal costing Islamabad $28 million. These aircraft came with extensive airframes, engine spares, and a simulator and more upgrading and overhaul work was conducted in 1992–93. The Bank of Credit and Commerce International (BCCI) was one source to have financed the deal. The transfer drew criticism from india who accused Australia of disturbing the regional military balance in South Asia. The Indian reaction indicated its systematic propaganda campaign aimed at maligning Pakistan and presenting Islamabad as the main culprit responsible for the imbalance. Thirty second-hand aircraft representing the technology of the 1960s were certainly no threat to New Delhi, who had acquired new MiG-29s, rated as a manifold better weapon system.

There were other decisions regarding the PAF's arms procurement as well, but with less strategic manifestation. For example, in 1990, a deal worth $250 million was signed for the procurement of 300 French Matra 'Mistral' missiles and launchers, obtained for both the Air Force and Navy. The Air Force said it needed these to defend the service's sensitive ground installations. The requirement was built around the service report on heightened threat perception in 1990. Air Chief Marshal Hakimullah, then the Air Force chief, had warned the government of his inability to defend the country if he was not allowed to buy the French missiles.[18] This is indeed intriguing because the Deputy Chief of Air Staff (Operations), who later served as Air Chief until early 1997, informed the author that there was no enhanced threat perception in 1990.[19] His comments were in response to a question about the general perception that India and Pakistan were on the brink of war that year. In any case, an independent Air Defense Command of the army had been established for securing sensitive ground installations and for providing overall ground-based air defense. It performed well during the Kargil operation in 1999 shooting down two IAF MiGs accused of violating Pakistan air space. Moreover, these missiles were not integrated into the 'Pakistan Air Defense System' (PADS) – an overall and complex communication and air defense system of the Air Force which was installed in 1977. One of the senior officers involved in the induction and integration of the system pointed out the discrepancy to the service's high command but was told to keep silent.[20]

Questions were raised about this deal, especially Prime Minister Benazir Bhutto's involvement in it. Allegedly, the brother of a close confidante of Bhutto, and some other members of her Pakistan People's Party, in collusion with the PM's husband, were involved in receiving kickbacks.[21] It must also be noted that during Bhutto's first government, her husband, Asif Ali Zardari, was accused of financially benefiting from his wife's position. Several contracts were cancelled to accommodate this particular procurement for which no trials were held prior to approval of the system. The involvement of PAF top management that played up the threat was certain. It must be noted that the central government does not have an independent source to confirm the threat situation presented by the military.

Arms procurement for the Army

After the PAF, the Army got the biggest share of American military assistance, electing for relatively cheaper options to increase its firepower, mobility, and technological strength. The main focus was to improve the armoured and artillery divisions of the service.

After the 1971 war, the main source of tank acquisition was China. In 1979–80 the Heavy Industries, Taxila (HIT) had started to overhaul and rebuild the Chinese T-59 tanks. These tanks presented the main strength of the armoured corps but they were, in fact, technologically inferior. American funding, however, was not used to obtain quality equipment by the army but to increase numbers. The service procured second-hand/refurbished but cheaper tanks such as American M-48A5s. These tanks being old and not in use by the US military were categorized as 'excess stock', which meant that they could be sold at 5–50 per cent reduced rate of the actual cost of the weapon depending on the condition of the hardware. The Congressional Presentation Report on the Security Assistance Programs for FY-1983 gave the acquisition value of 135 M-48A5s sold to Pakistan in 1981 at $78 000 a piece.[22] These tanks were preferred to the M-60s and M1A1s, which were more expensive. The procurement of the M-48A5s added to Pakistan's existing inventory of similar tanks received in the 1960s. There were subsequent plans to establish an overhaul and rebuild facility for the M-48A5s, and in 1986 a transfer of technology agreement was signed with the manufacturer, General Dynamics, but the project unfortunately stalled following the Pressler Amendment.

Islamabad declined an American offer for M1A1 tanks again in 1988–89. Despite Washington's willingness to sell its latest equipment that was in service with the US Army, the Pakistani GHQ decided not to buy them owing to that fact that they were more expensive and suffer from certain technical flaws. The tanks were formally rejected after the trials held in the desert area near Bahawalpur in August 1988. (General Zia-ul-Haq, who was then Army chief and President, had gone to Bahawalpur to witness the trials and took the fateful flight that crashed, killing him and 38 army personnel including five senior generals, the US Council-General in Pakistan, and another American officer on board.)

The primary choice remained the Chinese tanks that could add to the numbers. In 1987 the T-69s were procured with another order placed for the T-69IIs in 1989. Then, in 1990 Pakistan acquired about 200 T-85s. A senior MoD source claimed that Beijing had made these tanks on order from another country but, after the order was cancelled at the eleventh hour, Islamabad stepped in to buy them.[23] A contract was signed later for the transfer of technology for these tanks, designated as T-85IIP.

Such procurements fitted well with GHQs three-pronged strategy: (a) acquiring less costly but rugged Chinese tanks that can be overhauled, (b) assembling the Chinese tanks locally along with certain modification and upgrades and (c) manufacturing an indigenous tank. The process was initiated in 1979 when the first batch of T-59s was overhauled locally. This approach was also linked with the service's need to maintain three distinct levels in its tank inventory. The first would comprise the T-69s and the upgraded T-59s, followed by the second level of T-85s, and the third level consisting of the MBT-2000. The third and final part of the strategy was put into action in 1990. Under the leadership of General Mirza Aslam Baig, who took over as Army chief after Zia and who was ambitious about attaining self-reliance in weapon systems, a project was launched for the development and production of the MBT-2000 also known as 'Al-Khalid'. This tank was generally compared with tanks such as the Russian T-72 and American M1A1. (See Table 7.1 for comparative details of the three tanks.) The tank was named after a famous warrior from the days of the Prophet Mohammad. This not only expressed a level of sentimentality about history, but also represented General Baig's ambition to strengthen the armed forces and facilitate the formation of a strong Islamic bloc that could challenge the West, especially the US. Inheriting this mindset from his predecessor, General Zia, Baig was a great believer in attaining self-reliance and he had been a major force behind the Presidential directive of 1985 whereby all arms transfers were to be accompanied with transfer of technology to produce the systems in Pakistan.

Unfortunately, the plan did not materialize. Although the deadline of producing a prototype in 1991 was met, MBT-2000 never saw the production line. The failure was due to certain fundamental shortcomings in the local defense industry. Even though the tank did not start from the drawing board, the design having been based on the earlier designs of the Chinese T-69II and T-85 tanks, a tank takes time to develop. Not appreciating this reality was another example of how little defense industrial managers understood the dynamics production. Not only this, some officers of the army who were interviewed expressed their dissatisfaction with the service's capability to provide 'depot-level' maintenance facilities for MBT-2000. These sources were of the view that with the level of education of the average soldier, it would be difficult to repair the complex systems.

In time of war, it would be equally problematic to dispatch the tanks to the central depot for repair and maintenance.[24] Under the circumstances the Army may well have had to face the situation it had encountered in 1965, when it found it difficult to handle and operate

Table 7.1 Comparison of Pakistan's Main Battle Tank with Top Tanks of the World

	MBT-2000	T-72	M1A1
Country	Pakistan & China	USSR	USA
Crew	3	3	4
Combat weight (ton)	48	41	57.172
Power weight (hp/ton)	25–30	19	26.2
Length (longitudinal mm)	6900	7400	7918
Weight (turret top)	2300	2200	2438
Max road speed (km/h)	62	60	66.8
Main gun calibre (mm)	125 smooth bore	125 smooth bore	120 smooth bore
Muzzle velocity (m/s)	1760 850 905	1615 850 950	1650 1140
Auto loader	Yes	Yes	Yes
Ammunition storage round	39	40	40
Engine power	1200–1500	780	1500
Transmission armour projection	Auto-Hyde	Composite	Auto-Hyde
Cost	US$ 1.5–1.7 million	N/A	US$ 3.5–4.5 million

Source: Jane's All the World's Armoured.

American tanks to maximum advantage.[25] With delays in the MBT project, upgrading existing tanks was planned and, in 1989, various European companies were approached to facilitate the upgrade of M-48, T-54/55, T-59, and the T-69 tanks.

As part of the planning to enhance firepower and mobility, the Army acquired 24 Huey Cobra attack helicopters from the first American aid package (for details see Table 5.1). The 1965 war and the Indian military exercise Brasstacks, had proved that attack helicopters were needed for the anti-tank role in stopping the Indian onslaught. This procurement increased the service's defensive capabilities. In this case as well a comparatively cheaper option was selected. The Cobra was preferred to the more expensive AH-64 Apache which was offered by the US in 1990. In order to make the helicopters more effective, two significant categories of American anti-tank wire-guided missiles – TOW and the TOW 2 – were procured. Approximately 5000 missiles were purchased from 1983 to 1990. A contract was also signed in 1990 on behalf of the Pakistan Ordnance Factories with the American manufacturer, Hughes Aircraft, to produce TOW 2A anti-tank missiles in Pakistan. Again, the deal stalled after the Pressler Amendment was passed in October 1990.

One of the objectives of the GHQ at Rawalpindi was also to strengthen its infantry. Until the 1970s the infantry corps depended upon the G-3 and MP-5 rifles. Although the G-3 was a rugged item, more firepower was required which the army generals hoped to achieve by the procurement of various shoulder-fired missile systems. The most significant acquisition in this regard was the American FIM-92A 'Stingers'. These were acquired both directly and indirectly from the US. In the first instance, about 250 of these missiles were transferred to Pakistan from 1985 to 1987 and, in the second, more were siphoned from the 'Stingers' provided by the US to the Afghan *mujahideen*. Allegations were levelled against Pakistan in American Congress in which it was said that out of 600 missiles transferred to the *mujahideen*, only one-third had reached their destination and the rest were siphoned off.[26] It cannot be said with certainty that the Army retained all the stolen missiles, but the probability of a large number being inducted into the Pakistan Army was high. Some of these 'Stingers' also found their way onto the black market, where the Afghan *mujahideen* primarily sold them. Most of the stolen equipment was available for sale in Pakistan, mainly in the two provinces bordering Afghanistan: NWFP and Baluchistan. The Army's involvement in diverting the weapons fanned a particular opinion regarding the involvement of certain generals in the explosion at the ammunition depot, 'Ojhri', near Rawalpindi. The explosion took place at a time when questions were being asked in Washington about the 'Stingers', and allegedly there were plans of an American team visiting Pakistan to take account of these missiles. The damage to human life was great in this accident.

Other shoulder-fired missiles such as the Swedish RBS-70 were also acquired. In 1986, a total of 84 RBS-70 launchers and a hundred missiles were imported from Sweden as Completely-Built-Units (CBU). It was planned to start their production locally at a later stage. These missiles were integrated for use with the Swedish fire control radar 'Giraffe' procured for assembly at Margalla Electronics in 1985.

A project for the manufacture of indigenous shoulder-fired missiles was also launched. The 'Anza' and 'Anza II' were made with Chinese help. The official sources made claims to the effect that the two missile types, which were developed under Dr Abdul Qadeer Kahn's supervision, were a symbol of the nation's capability in missile manufacturing. These two missile systems, however, primarily consisted of Chinese subsystems that were assembled in Pakistan. A similar publicity campaign was launched about the ballistic missile project that apparently led to the production of two systems: 'Hatf-I' (range: 80 km), and 'Hatf-II' (range: 300 km). These were developed and test-fired in 1989 in response to India's missile and space technology program. Islamabad had wanted to acquire ballistic missile capability and increase options for a nuclear delivery system for some time. In 1986, the authorities signed a $40 million deal to covertly obtain know-how to develop ballistic missile capability under a project code-named 'Khyber Pass'. Although the plan was foiled, it demonstrated Islamabad's covert activities related to the procurement of sensitive materials and information for its nuclear program.

The manufacture of 'Hatf-I' and 'Hatf-II' was Pakistan's response to the ballistic missile competition in South Asia. Experts, however, were not very convinced of Pakistan's capacity to match the Indian program. This analysis was based on various reports on the performance of these missiles and the knowledge that the missiles were predominantly based on Chinese materials and technology. Approximately 80 per cent of both systems were Chinese, with Pakistan's primary input being the assembly work. A greater Indian and international concern was about Islamabad's acquisition of the Chinese M-11 missiles with a range of 300 km, upgraded to 600 km. Both governments refused to disclose any information or the exact number of the missiles transferred. The M-11 was Pakistan's greatest defense against India's 'Prithvi' and 'Agni' missiles. The sharpest reaction was from the US government, threatening both the Pakistani and Chinese governments with sanctions if the transaction was carried out. Washington's main concern was to stop a ballistic missile race in the region and to discourage the transfer of technology that would enable any regional state to deliver nuclear warheads. The US propaganda, it was felt, was aimed at Pakistan alone. This was basically because Islamabad's dependence on American military and economic aid provided Washington with the leverage to twist Pakistan's arm diplomatically.

Despite official commitment to indigenous production, Islamabad failed to strengthen its defense industrial base and related self-reliance during this period. The excessive dependence upon American source of procurement, therefore, was natural. This dependence could not be reduced with the kind of poor planning being carried out. For instance, Margalla Electronics was established in 1985 for the local manufacture of ground-based radar. The infrastructure was completed in 1987 to start work on American LAADS radar. A total of nine LAADS radar worth $1.06 million per unit were in CBU form. Another two were procured for $1 million per unit in semi-knockdown (SKD) form. The final delivery made was of four CKD kits worth $0.95 million per unit and the LAADS radar project was finally completed in November 1991. In 1987 another order was placed for 14 Italian 'Skyguard' radar. These radar, which came in SKD form, cost $2.58 million and were financed through the US foreign military sales program. In 1986, 17 'Giraffe' (fire control radar) were ordered, consisted of four CBUs for a price of $0.84 million per unit, four SKDs (to be assembled in Sweden), four SKDs (to be assembled at Margalla Electronics) with per unit cost of $0.8 million, and five CKDs costing $0.79 million per unit. The objective was that Margalla Electronics should attain 70 per cent deletion of foreign components by 1994 but it failed to meet the required goal, which is not surprising considering two factors: (a) industrial and technological discrepancies in the field of electronics, and (b) a strong import lobby that preferred importing equipment including radar rather than anchoring the manufacturing technology locally.

The creation of Margalla Electronics was MoD's idea for catering for the three services, although it originated from the PAF, which wanted an independent radar manufacturing facility for itself. Ironically, the idea was not endorsed by any of the services. The Air Force did not want to go for a joint project, and the other two services by then had not formulated any plans to manufacture radar locally. The Army, particularly, was fulfilling its needs by procuring fire control radar from the US. MoD, furthermore, toyed with yet another idea of a facility to make printed circuit boards. The services again did not show any interest since all had independent facilities to manufacture printed circuit boards. No one involved in the decision made any effort to reverse the process of creating such a facility. On the contrary, people responsible for making the decision tried to justify its creation by opting for the original idea – a radar-manufacturing establishment. Was the facility really needed when the Air Force had planned to set up a radar-manufacturing unit at PAC? At the time Margalla Electronics was established the Air Force's case was already under consideration! The factory at Kamra was set up in 1987 to manufacture low-tech ground-based radar. The only explanation for the duplication lay in the competitive environment within the military bureaucracy.

Equally intriguing was the establishment of the Institute of Optronics in 1985 for the manufacture of night vision devices/goggles. Work started in 1986 for assembling four different types of devices, and the planners hoped to attain 30 per cent deletion of foreign components. The first delivery to the Army was in 1988. In the early 1990s it was claimed that 40 per cent deletion had been attained[27] but the calculation was made on the basis of the number of components. Even with this principle the deletion figures appeared exaggerated, not to mention the fact that any deletion attained by IOP was limited to mechanical components and not the electronic or optro-electronic parts. Merely the testing of imported tubes and lenses was carried out regarding this part of the engineering making any official claims debatable. The circumstances under which the facility was established were dubious as well. The facility was headed by Dr Abidi who, despite being a civilian, was given *carte blanche* by the then Army CGS, Lt General Aslam Baig. One Pakistani source was of the view that Abidi sold the idea to the over-enthusiastic General Baig.[28] Abidi not only got the materials to manufacture the night vision devices and goggles from the US, but also managed to procure human and industrial resources from another R&D establishment, DESTO. The chief scientist at DESTO claimed that a major part of IOP's infrastructure actually belonged to DESTO.[29] What is even more interesting is the fact that after this diversion the government reinvested in acquiring new equipment for DESTO. The products produced by IOP could not satisfy the infantry divisions, who complained about the systems' ineffectiveness.

There was a lot of publicity about the indigenous manufacture of the LRF 786P laser range finder claimed to be made by Dr Abdul Qadeer Khan's team. Although there were no details available to assess the equipment, sources were of the view that the laser range finder was not purely a Pakistani product as people were led to believe. It was assembled from imported components mainly from China.[30] The indigenous range finder, actually produced indigenously by another organization, was pushed aside after Dr Khan presented his equipment, which he claimed was made locally at his laboratories.

Other acquisitions for the Army comprised howitzer guns and AN/TPQ-36/37 ground radar from the US. Moreover, in 1984 a deal was signed with China for establishing a new machine-gun factory. In 1989, another project was launched to manufacture trucks. The Yasoob truck project was a joint public–private sector venture to produce 1250 6 × 6 (weighing 16–22 tons) and 1750 4 × 4 trucks (weighing 12–15 tons) with a per unit production cost of $40 000 and $29 000 respectively. These trucks were to be added to the Army's inventory but, in fact, one did not find many Yasoob trucks in the service! In the view of Maj. General Salimullah, this shortfall was related to the inefficiency and poor business practices of the manufacturing firm. He told the author that the company did not meet its commitment despite having been provided with the resources in advance.[31]

Arms procurement for the Navy

The Navy was not able to procure any major weapon systems from the first aid package. By 1979, the gap between Pakistan and India's naval capabilities had grown. The difference in terms of manpower and equipment was 1:4 and 1:2 respectively. Matching the Indian Navy's 'blue-water' capability was never a priority of Army generals whose views were fundamental in setting arms procurement agendas. Despite this bias the Navy had done quite well for itself in terms of acquiring smaller systems. In the 1970s the service had obtained an aviation wing and a few French submarines. In 1980 six Gearing class frigates worth $41 million were procured on lease from the US as well as three sets of Harpoon missiles and launchers procured for $156 million. The amount was minimal compared with what was spent on the other two services.

This situation continued even after Washington offered the first military assistance program. The Reagan administration was not interested in an overall military modernization of its ally, especially the Navy. The idea was to strengthen Pakistan to a degree that would convince Moscow of Washington's commitment to check USSR's expansionist designs and no more. Equipping the Pakistan Navy was certainly not on the agenda. Besides, until the mid-1980s, America had nothing to offer to the Pakistan Navy as there were no naval vessels that could be transferred to Pakistan as part of the US Navy's decommissioning exercise. Procuring new but expensive American equipment was not affordable.

The change came in the second half of the 1980s when eight frigates were offered on lease to Pakistan. The transfer comprised four Brooke class and four Garcia class frigates. These ships were part of an American plan to decommission around 600 ships from its Navy. The apparent change in Pakistani decision-makers' attitude towards the Navy did not imply any shifts in the strategic planning. After the potential benefit of the first aid package drifted away, the Navy adopted a course of publicizing its strategic needs and pleading its case through the national print media. The public was appraised of the growing threat of India's naval capabilities and Pakistan's own weaknesses in guarding the SLOCS and defending its only seaport, Karachi. Not that public opinion has ever played a major role in defense decision-making, but the media was used to conveying the junior service's frustration to the top policy-makers. It was also a matter of General Zia keeping the other two services happy by giving them some of what they wanted. In any case, the deal was financially viable with the eight frigates costing Islamabad US$9 million with an additional $186 million for the acquisition of armaments and other support equipment. This included three SH-2F Seasprite anti-submarine helicopters, 64 Standard MR-SM1 anti-aircraft, and 64 Honeywell MK 46 MOD 5 light weight anti-submarine torpedoes.

For Washington it was beneficial to strengthen the Pakistan Navy at that juncture with the ultimate objective of utilizing its capacities for the security of the Persian Gulf.[32] In 1988 a deal was signed for the transfer of three P-3C II.5 version aircraft. Washington needed a repair facility for its own P-3Cs visiting the Arabian Sea littoral. Earlier during the Afghan crisis Pakistan provided transit facilities to American P-3C Orions, and the US had plans to establish a logistics base at Karachi to provide a maintenance facility for its aircraft operating in the region.[33] This contract worth $214.6 million was to be paid from the second military assistance program. The deal included spare support of 'O' and 'I' levels, spare engines, documentation and publication, contractor engineering technical services (CETS), training, software support, ground support facility, logistics technical assistance and program management. No planning was done at that time regarding depot level maintenance. It was found too expensive for the PN to acquire a depot level maintenance capability for a limited number of aircraft. The first aircraft was to be transferred to the Pakistan Navy in April 1991 with the other two following in July and October 1991. The aircraft had reconnaissance and anti-submarine warfare (ASW) capability, which provided Pakistan with a limited surveillance capacity at sea.

The Pakistan Navy had considered the replacement of its old ASW-capable French Atlantic-1 aircraft. A Pakistani source revealed that these older aircraft were being maintained by acquiring spares from India via third country.[34] It is possible that spares were procured through private defense contractors who might obtain them from India without the knowledge of the Pakistani authorities. Although this factor alone would have negative implications at the time of war, it did not, however, play a role in the Navy's buying a new type of aircraft. In fact, beefing up the aviation wing was not one of the priorities at NHQ where procurement planning is dominated either by submariners or surface ship operators. There was definitely no staff requirement at the service headquarters for equipment comparable with a P-3C or, for that matter, any other aircraft.

The American aircraft was selected because at that time there was no other equipment that the PN could buy from the US. It did not, however, want to surrender the resources to its sister services. There were a number of senior officers who, then, had disagreed with the selection on the grounds of a potential arms embargo.[35] In 1987–88 there were signs that American policy towards Pakistan might change after a solution of the Afghan crisis was worked out with Moscow but such comments were deliberately ignored. One of the factors was that decisions in the Navy, like in the Air Force, were dominated and forcefully directed by the top management. Senior officers' interference was guaranteed in an environment where officers dealing with particular procurement are posted out without leaving sufficient paperwork for the successor to follow the trail. There is no system of project management; and the frequently changed officers, who have little notion of the case history, would often resort to simply obeying directions passed on from above without presenting their own analysis. The procurement of the P-3C was one glaring example of the ineptness of naval staff. The deal, however, stalled in October 1990 owing to the imposition of the arms embargo. Interestingly, payment to the US OEM due to be completed in 1993 was never stopped.

Not being able to get American equipment the Navy looked at other options. The natural tendency was to look at British hardware that was familiar to the service. As a result, two ageing Leander class frigates and a County class frigate were purchased from the UK. These ships were obtained at a throw-away price ($20 million for the two Leander class and $3 million for the County class), but they were fairly old and the Royal Navy got a good bargain in selling these ships to Pakistan. Meanwhile, negotiations were also carried out for the procurement of new ships such as the Type-21 and the Type-23. British naval officials were invited to discuss in detail the modernization plans for the Pakistan Navy as well as the prospects for refitting the old frigates. The team encouraged PN officials to consider the Type-23, but negotiations lasted for more then nine years mainly because of the PN's inability to decide whether it wanted Type-21 or Type-23. The talks initiated around 1982–83, for the procurement of three or four Type-21 frigates, failed for three reasons: first, the British government did not agree to certain modifications demanded by Islamabad. Secondly, the Ministry of Finance was unable to generate the required resources to finance the deal. At that time the country's meagre resources were committed to finance Prime Minister Junejo's five-point development program. Democracy was reintroduced after eight years and President Zia was under pressure to help the new regime function, or at least give the impression to that effect. Sacrificing naval procurement was a viable option. Needless to say, the Navy did not have a strong enough lobby to obtain resources from the democratically elected government, or to convince General Zia to persuade the DCC to approve procurement of British frigates.

Thirdly, any prospective deal was stalled owing to the naval chief, Iftikhar Sirohey's bias against the purchase of Type-21s. He was inclined towards the bigger and advanced Type-23. This peculiar bias caused the Navy to turn down an offer for the new Type-21s to be accompanied by the transfer of technology from Britain to manufacture at least one ship in Pakistan. Sirohey also turned down a Turkish offer for three frigates at a price equal to one British frigate. A senior source from the service affirmed that Sirohey was promised the position of the chief of the service if he did not approve the purchase of the Type-21s.[36] This case illustrated the decision-making culture in the service that, compared with the other two services, was dominated by its top management with little concern for the views of equipment operators or end-users. Certainly in this case it all depended upon the will of the naval chief. The British government was persuaded to revise the offer and agreed to consider the transfer of Type-23s. The new British offer came in 1987 for the Type-23s, which were to be fitted with weapon systems and sensors used by the Royal Navy. It was the first time that London had agreed to sell ships and systems in service with its Navy. The deal progressed to a stage when a Cabinet subcommittee appointed by Prime Minister Junejo approved the deal, despite opposition from certain parliamentarians. The prospective deal apparently fell through because of lack of funds and coincided with the time when the government started to worry about the cracks that had been appearing in the national economy since the late 1970s. Since that time, the country's external debt burden had increased to $20 billion.

Admiral Sirohey's fascination with the latest technology made him look at the possibilities of procuring a nuclear submarine. The quest for this equipment followed India's acquisition of a 'Charlie' class nuclear submarine on lease from the former USSR. Induction of nuclear submarines would increase India's chances to block Karachi or strangulate its adversary's Navy at sea. Since nuclear submarines are less noisy than the diesel-powered and more difficult to detect underwater, they add to the Navy's ability to surprise. It would, however, require more then one nuclear submarine to become a potent naval force, and Pakistan was definitely not in a financial position to acquire one single nuclear submarine, let alone an entire fleet, but still Sirohey went ahead planning the acquisition. In 1988 the Canadians were approached for its conventional nuclear hybrid submarine and in 1990 Islamabad looked at other sources that could sell nuclear subs. When asked by the author if the Pakistan Navy had the capacity to operate this technology, Sirohey responded that he had plans to train over 200 personnel to manage the repair and maintenance of a nuclear submarine. He added that, had General Zia not died and he himself not been made the Chairman JCSC, he would have managed to procure a naval submarine.[37] The former Naval Chief's opinion not only displayed his earnest efforts to introduce modern technology to the service, but it also highlighted major lacuna in the entire procurement policy-making process. The JCSC has no role to play in defining the overall strategic planning of the country or the authority to prepare an integrated priority list for the weapons most essential for military security. Was the idea dropped because procuring a nuclear submarine was not part of the national strategic plans? Or were the decision-makers within the military and the government unconvinced of the value of acquiring such technology? The capital expenditure and life-cycle cost of a nuclear submarine being exorbitant was not affordable without compromising procurement plans for the other two services. Such expensive shopping was strategically meaningless because the acquisition of a single nuclear submarine would not have contributed significantly to national defense. Although the submarine was not acquired, the example shows how some decisions were driven by the service's top management's desire to equip the Navy with modern weapons without first consulting the overall strategic or war plans.

The decade of the 1980s was marked by the Navy's assertiveness in trying to compete with its sister services in attaining an organizational strength, at least at par with the Air Force if not the Army. Two examples of this relate to the procurement of the French Matra 'Mistral' missiles and creation of a new division of marines. In 1990 when the PAF had demanded these missiles for the security of sensitive ground installations the Navy also presented a similar demand. This procurement was wasteful for the same reasons as the PAF's acquisitions of the 'Mistrals' were considered objectionable. It is not known which of the Navy's agencies was to be equipped with these missiles. By 1990 the Navy already had two independent sub-organizations for coastal defense: the Navy's Special Service Group (SSG) and the Army's division dedicated to the defense of the coastal facilities. In addition, a division of marines was created in 1989–90, although there was no information available as to why the Navy required three different agencies to guard its limited number of coastal facilities.

Other less significant acquisitions made by the service pertained to four Huangfen class missile attack crafts. These were transferred from Beijing in 1984 for about $20 million per piece and were later modified and improved to suit the requirements of the Pakistan Navy. It must be remembered that the Navy has never been keen on obtaining Chinese equipment that is considered qualitatively inferior. These missile crafts were obtained mainly because of the low cost as well as the possibility of striking a deal of technology transfer for manufacturing them locally at a later stage.

The naval acquisitions that began towards the end of the 1980s were to continue into the 1990s. Importantly, these signified the level of inter-service competition that resulted in wastage of resources. Financial mismanagement was observed in the acquisitions related to indigenous defense production programs, some projects for naval procurements, and purchases of less significant weapon systems. All of these were financed through national resources, which leads to the conclusion that the PN has a tendency of wasting resources.

8
Military Buildup Decisions, 1990–99

During these ten years military buildup depended upon national resources. The introduction of the Pressler Amendment to the US foreign assistance act blocked arms transfers to Pakistan in October 1990. The vital features of this period were naval arms acquisitions since national resources were not sufficient to cater for conventional military modernization. The state of the economy and political instability added to deteriorating financial conditions. It was, thus, that dependence upon non-conventional defense grew to a degree that policy-makers started to consider the nuclear option as a panacea for all military strategic and technological problems.

Arms procurement for the Navy

In these ten years the biggest share in major equipment procurement was received by the Navy, which has the least significant role in military-strategic plans. This was achieved through consistent publicity of the service's requirements. A series of articles was brought out about the service's responsibility of guarding naval trade routes, although this argument did not take into account the fact that only five to ten per cent of the country's trade was carried out by its own merchant navy. The Pakistan Navy's strategic planning had before never gone beyond defense of the SLOCS – a limited role natural for a small navy. A prospective Indian blockade of the SLOCS and the only sea outlet at Karachi could completely paralyse the country especially in a protracted war. Earlier, in 1971, the Indian Navy had blockaded Karachi, and this threat was close to becoming a reality again during the height of India–Pakistan military exchanges at Kargil. Pushed to the wall, the Indian Navy had plans to blockade Karachi.

The service's approach towards strategic planning was markedly different from its sister services. The Army in particular looked at a potential medium intensity conflict as a short duration war, possibly limited to a week or ten days. Contrastingly, the Navy's top brass was of the view that a fourth war between the neighbors would be long with India holding Pakistan under siege through the sea. The Commander Pakistan Fleet in 1994, Admiral Shamoun Alam, was of the view that in order to avoid this situation the service needed a minimum of 12 submarines and over 24 ships.[1] In 1999 the Navy had a fleet of about six submarines and eight frigates. Army generals were not keen to divert resources to a smaller sea-based service when it was difficult to comprehend the strategic significance of the navy in a country not entirely dependent upon sea trade for its survival. What the GHQ failed to realize was that, with relations turning sour with Tehran and chaos in Afghanistan, Pakistan required the safety of its sea routes of supply to survive through a war. Increasing the Navy's share of the defense budget required financial sacrifice by the Army that it was not ready to give. Despite this, the Navy managed to get some equipment in the 1990s. In fact, during this period it was the only service to get four major deals through.

The first deal was concluded on 17 January, 1992 for three French mine hunters worth $350 million. This included the transfer of the technology for one mine hunter in SKD form to be assembled in Pakistan. Although some sources challenged the cost effectiveness of assembling one mine hunter in the country, the deal was signed in compliance with the presidential directive of 1985. The naval officers, conscious of the relatively high cost of the transfer of technology, argued that this would enhance the potential for naval equipment manufacture. Islamabad paid 20 per cent of the total amount as down payment, the French government providing the credit with a six to seven per cent rate of interest. Pakistan had to pay about $25 million in six-monthly instalments, the final payment to be completed by 1997. The net worth of the deal was approximately $500 million. These mine hunters were needed to counter the possible threat posed by India that possessed over 2000 mines. The top management that replaced the naval chief, Admiral Mansoor-ul-Haq, and his team in 1997 was of the view that there was no real threat of Indian mine-laying operations and the mine hunters were not required. Interestingly, the new management was more concerned about the high-handed attitude of the US Navy that had fired 'Cruise' missiles on Afghanistan in 1999 from Pakistan's territorial waters. A plan was also discussed to dispose of these vessels to a friendly country in the Gulf. There were definite flaws in the strategic weapons requirements owing to the absence of a streamlined procurement system linking weapons acquisition planning with threat assessment.

Another deal with the French was signed in 1994 for the acquisition of three Agosta 90-B submarines worth $950 million, for which a loan was provided by the French to be paid in five to six years. Hence, the net amount payable to the French was $1.3 billion. The first submarine to be completely manufactured by the French was delivered in September 1999 while the second, jointly manufactured by French and Pakistani technicians at Toulouse in France, was released for transfer in November 1999. This submarine was withheld for some time because of the temporary embargo imposed by Paris in October of the same year. The third submarine was being manufactured at Karachi and was to be completed by the year 2000. An additional attraction was Paris's pledge not to sell these types of submarines to India, as well as an agreement to assemble one submarine in Pakistan that would be provided in CKD form with transfer of technology in certain key areas. This will be added to the existing inventory of six submarines and would go some way to improving the Navy's defensive capability against India's powerful naval force. These submarines could deter the Indian Navy from efficiently blocking Karachi, or landing its troops there. In addition, these submarines with an air independent propulsion system (AIP) could remain submerged for longer making them more difficult to detect. The approval for procuring the submarines was obtained from the DCC in 1990–91 during Prime Minister Nawaz Sharif's first government. This was considered the next best option after nuclear submarines, which the navy was still considering in the early 1990s.

Negotiations were conducted with three sources: France, Sweden and China. The Swedes had offered their Type-96 submarines with AIP technology. The Chinese also made an offer to sell their subs for approximately $180 million, $83 million less than the French offer. French subs were $70 million cheaper than these of Sweden. Despite the fact that the Chinese deal would have given Islamabad the advantage of acquiring a greater number of submarines, the offer was not entertained because of the Navy's bias for West European equipment. The final decision, in favor of the French submarines, was driven by the service's need to acquire the best technology available in order to meet the technological shortfall with India. Although based on strategic logic, the argument did not take into consideration the fact that, even with the qualitative enhancement gained through the acquisition of these submarines, the Navy would find it difficult to counter India's both qualitative and quantitative superiority. Needless to say, the deal had major lacunas. For instance, the contract did not include torpedoes that were necessary to defend against an attack by enemy submarines and surface ships. The excuse for not including the torpedo in the main deal was that, at the time, the French F-17 MOD 2B designed for this submarine was still under development. This was actually used as an excuse to break up the procurement into more then one contract, a practice that usually serves the personal interests of those who engage in negotiating kickbacks. It was in 1997, when plans were being put into place for the transfer of the first submarine to Pakistan, that the service started to look for ways to arm it. Interestingly, the service dropped the idea of acquiring French torpedoes and showed interest in the Italian A-184, which was cheaper than the F-17 MOD 2B. The price quoted for 72 A-184 torpedoes was $80 million, whereas the French package was for $96 million. In 1998 a request by the service to the government to provide $100 million to procure torpedoes was turned down for lack of funds.

In 1992–93 another deal signed related to the procurement of six second-hand British Type-21 frigates for $150 million. The transfer begun in 1993 concluded in 1995. The deal was totally self-financed and included spares and training but excluded the Lynx helicopters that the Royal Navy used on these frigates. Even in this case, the procurement deal was broken up into two. The AWS-capable Lynx helicopters were not part of the main contract but were acquired later through a separate deal in 1994. As mentioned earlier these were old ships that the Royal Navy had planned to decommission. With about 2300 hours remaining for the engines of these frigates, talks were conducted with the British authorities to transfer the technology for their overhaul. The British government, however, did not agree to provide this capability. It must be remembered that when negotiations for these frigates first began in the 1980s, the British government had offered to sell new Type-21s with the transfer of technology. The Pakistan Navy, nevertheless, bought these old ships mainly owing to its need to enhance its frigate fleet at minimal cost: the new Type-21s with transfer of technology would have cost Islamabad much more. There was definite disagreement as to the choice of these ships, some officers favouring the purchase of British Leander class frigates, as these were part of the PN's existing inventory, and two of the Type-21s had damaged hulls. The naval chief, Admiral Mansoor-ul-Haq, who instructed all his officers to support a decision for these frigates 'with one voice', forced a final decision. Officers still opposing the selection were conveniently side-tracked. The selection was justified on technical grounds: this acquisition was necessary to fill the gap created by the withdrawal of eight American frigates in 1989–90. These ships could supposedly help (a) in creating the capacity to endure and repulse an attack on the single naval base/port, Karachi, for at least eight to ten days, and (b) to complete the Navy's chain of defence.

Why did the Navy suddenly get all this attention? Was it because of some changes in strategic planning whereby it was considered necessary to enhance the Navy's defensive capabilities? Did the elected leadership start to understand the need to secure the SLOCS and EEZ? These deals, nevertheless, did not denote any change in strategy, nor did the political leadership exhibit any vision regarding deployment of its armed forces. After 1985 the elected governments were too preoccupied struggling for political survival to contemplate strategic issues that were basically the forte of the armed forces. One of the significant explanations could be the financial kickbacks. Apparently, approximately $120 million went into the pockets of top decision-makers in the Navy and the civilian government. In 1997 the naval chief, Admiral Mansoor-ul-Haq, was sacked on corruption charges related to the French Agosta 90-B submarines deal. Bhutto's second government approved the purchase and Sharif's subsequent government accused her and her husband of taking bribes for approving this deal. Dr Mehboob-ul-Haq an eminent economist, repeated these allegations as well. Benazir Bhutto never contested the claims regarding her involvement but, in a letter written to the chief of the accountability cell in January 1998, Naseerullah Babur, one of Benazir's aids, accused Shahbaz Sharif, the Punjab Chief Minister and the younger brother of Prime Minister Nawaz Sharif, of being involved in the deal. It was difficult to trace the number of people involved in receiving kickbacks because the deal was handled by three governments: (a) Nawaz Sharif's first government, (b) the caretaker government in 1993, and (c) Benazir Bhutto's second government. What is certain was the involvement of senior naval officials who delayed the negotiations during the tenure of all these regimes. The procurement, moreover, was made possible by Mansoor-ul-Haq's assertiveness and personal interest. Haq was highly motivated to strengthen the Navy as an organization and it was this drive that made it possible for the service to acquire the submarines that the French had been offering since the 1980s. These were offered for the first time during Zia's presidency but were not considered due to his view about a military strategy that did not assign an important role for the navy. Not that the Pakistan Navy did not need submarines, but the particular choice of relatively expensive French submarines was possible because of Haq's maneuvering and his persuasion of the government.

The transfer of technology deals did not benefit the service either. Senior naval officers claimed that the French backtracked in providing Islamabad with certain sensitive technologies that they had agreed to in the contract. In July–August 1999, in the wake of a new round of India–Pakistan hostilities, Paris also imposed a temporary embargo. Again, the embargo was re-imposed after Pervaiz Musharaf's coup in October. Consequently, no substantial progress was made in naval production despite the high cost of the deal. The total contract value was $1.3 billion, which excludes the resources spent on establishing the Submarine Re-Build Complex at the new naval base at Ormara.

In 1994 a collaborative venture was initiated with the Chinese for the manufacture of gun and missile boats. Initially, the idea was to manufacture gunboats but after manufacturing six boats of this type, it was decided by the service's high command to switch over to missile boats. The construction was carried out stimultaneously at the Naval Dockyard and KSEW, the Chinese providing the technology for the construction of the platforms, but the boats were fitted with components from the West. The Navy high command linked this project and that of the French mine hunter and submarine assembly as part of a uniform policy to encourage indigenous manufacture of equipment. It is difficult to ascertain how the Navy top brass hoped to attain their objective with the assembly of a limited amount of equipment, and without developing the relevant technical and industrial know-how in the country.

Negotiations on procurement of the Chinese F-22 frigates with transfer of technology were also resumed in 1997–98. The initial talks for the F-22 had begun in 1985. There were people at the service headquarters who had opposed these frigates, preferring the South Korean HDF-2000 made by Hyundai. However, the naval staff requirements finally made in the early 1990s were to fit the Chinese F-22 frigates only. A blanket approval for the Chinese vessels was obtained in 1992–93 at a cost of $700–800 million. This included the cost of technology transfer and upgrading of the KSEW, where these frigates were to be constructed. The ultimate objective was to enable Pakistan to manufacture these naval vessels for future exports. Similar plans were also made for indigenous construction and export of the midget submarines of Italian origin. The midgets in the PN's inventory were purchased in the 1970s. These had undergone an overhaul in 1997–98 in Karachi giving the top management of the service the idea that they could easily manufacture these locally. Moreover, it was believed that the midgets could be sold to the navies of some countries of the Gulf and the Far East but it was difficult to see how the PN could manage such a venture. A final deal was stalled owing to the inability to negotiate a favourable price and disagreement within service headquarters regarding the technology concerned. There were problems with the signatures and the hull of the frigates. The Thai navy, which had also purchased these vessels, had to refit and overhaul their hulls to eradicate the problems. There were a number of people in the NHQ who supported the idea of considering other suppliers such as South Korea and Ukraine. The top management's view was, however, that including other options as late as 1998–99 would force the DCC to reopen the case giving an opportunity to the other two services to turn down the procurement altogether. The DCC had approved the procurement of three to four frigates in 1994–95 but selection was not finalized then. It was feared that reopening the case would allow other suppliers to influence policy-makers. The top management's attitude towards possible diversification of sources of procurement was found rather peculiar. In an interview with the author, the naval chief, Admiral Mirza, then the Vice-Chief, ruled out the idea of the PN approaching East European manufacturers. His main argument was that if these sources were interested in selling they should initiate a contact. He further expressed his scepticism about East European manufacturers' ability to provide spares support.[2] The naval chief's view depicted a certain naivety of the weapons procurement business, or was it that he did not want to speak about other pressures? For instance, during the frigate talks with Beijing pressure was exerted directly by the Chinese government on Islamabad to sign a deal. This pressure included financial kickbacks.

A deal could not, however, be finalized because of the paucity of funds. In 1999, the government was in no shape to allow major weapons acquisitions. This situation also put the service's modernization plans into disarray, including the idea to decommission the six Type-21 frigates that the service had planned to retire by 2010. This plan could only be sustained if the service could get new frigates from China or some other source. There were problems with maintaining the Type-21s owing to scarcity of spares because some of the US-manufactured components could not be obtained by Pakistan after the arms embargo. In 1998–99 there were plans to cannibalize one ship to keep the other five afloat. The question is, why did the PN procure frigates which used American components? Officers at the service headquarters were of the view that the ISI report on Pakistan–UK relations had not forewarned the service of any such problems. The ISI being the main source of input on external affairs, its information was never challenged.[3] The Foreign Office was not consulted at all. There were officers who supported the procurement of Leander class frigates that used British components and were already being operated by the PN. Furthermore, there were sufficient spares stocks to maintain additional numbers of these vessels. The suggestion, none the less, was overruled.

At the end of the 1990s, the service tried enhancing naval R&D. The Commander Logistics (COMLOG) established NRDA in 1998. The ACNS (Tech) would provide the interface at NHQ. There was tremendous resistance regarding the control of the organization by the Plans division that wanted NRDA to be controlled by it, and the branch rivalry in the Navy was intense. The Plans and Operations branch wanted control of every single major activity including manufacture and sale of equipment. This has a negative impact on the performance of NRDA. The Plans division also insisted upon controlling various arms exports negotiations. Talks were held with Qatar and Malaysia to assemble midget submarines and the submersible 'Chariot' mod. CE2F/X100-T of Italian origin with assistance of the OEM. The idea was that an Italian company, COS.MO.S, would provide the 'kit of material' of the equipment to be assembled in Pakistan. The PN would provide cheap labor for assembly, training and warranty to the end-user. The draft proposal for this joint venture prepared by the OEM lay the responsibility of covering the warranty cost with the PN. A similar idea was floated for assembly and sale of midget subs of Italian origin. These projects could not take off because PN officers did not understand the intricacies of defense sales. Finally the OEM struck an independent deal with one of the Malaysian shipyards totally side-tracking the PN that had great designs for arms exports.

It was also in 1998–99 that the PN started looking for buyers for its Lynx helicopters. It could no longer operate these because components were scarce. The helicopters were part of the $600 million inventory of excess stock the PN wanted to dispose of. An idea for establishing a defense sales and export organization on the pattern of the British DESO was floated for the purpose. In early 1999 the Prime Minister Nawaz Sharif gave verbal approval of the plan to NHQ to establish such an organization for the three services. The idea, however, was hijacked by the Army that envisioned the establishment of a defense sales organization controlled by JSHQ. By the end of 1999 JCSC was an organization controlled by the Army. Interestingly, no questions were ever raised about how the service had come about stockpiling such a volume. This also depicted the lack of cost consciousness in military personnel. It was generally not appreciated that the high wastage would be consequential for future allocations and, in turn, determining the role of the service. In the armed forces, particularly the Navy, wastage is pretty high. Funds are wasted through lack of knowledge of contracting especially when it comes to the rate-running contracts. None of the three services carries out cost accounting of its inventory, projects or other activities.

The PN also had problems operating the P-3C Orions, released by the Clinton administration in 1997–98. The reactivation program for these aircraft cost $4.028 million. However, the three aircraft were transferred in a 'fit-to-fly' condition and not in a 'fit-to-operate' condition. A number of critical spares required to meet AoG conditions were not provided. Having little knowledge of these aircraft PN officers didn't realize that a large number of the spares supplied by the OEM prior to the embargo were short shelf-life items. Sources other than the OEM were sought to negotiate spares support and crew training. Regular flying, let alone operational flying, could not be carried out without getting fresh crew trained. Interestingly, most of the crewmembers, including pilots, taccos and navigators, had retired by 1998–99. While dispatching the crew for training in 1989–90 the age bracket of the personnel was not considered. The training had not been completed when one of the P-3Cs crashed during the annual naval exercises in October 1999.

A general notion was that the procurement of the P-3Cs would increase the enemy's cost of defense. These aircraft with their superior capability could reach as far as the Indian naval port of Vishakapatnam and it was believed that, to counter the threat posed by operations of these aircraft, the Indian Navy would have to re-deploy its existing forces. The service's annual strategic exercise, 'Hammerhead', however, did not evaluate the cost of sending three P-3Cs, basically reconnaissance and anti-submarine warfare capable aircraft, deep into the enemy's area of operation. There was a disparity between calculations presented by planners and operators for the number of flying hours required for operational readiness of the aircraft. The Navy did not have fighter aircraft, nor could the PAF spare an entire squadron of its F-16s in case of a war with India. Indubitably, the PN's management alone could be blamed for such a blatant lack of analysis.

Arms procurement for the Air Force

The US arms embargo left the PAF to struggle with the growing gap between its own and the IAF's technological capabilities. In 1997 New Delhi placed another order for the Russian Su-30 fighter aircraft that were to provide it with the capability to target all Pakistani cities. The Russian aircraft were to replace the aging fleet of Russian fighter aircraft and offset the IAF's technological problems. None the less, it added to the concern of the PAF's top brass that was equally concerned about its own fleet. Training on the F-16s was limited because of shortage of spares. The service had to depend upon short endurance Chinese aircraft such as F-7s. The PAF was extremely conscious of the situation. According to Lt General (Retd.) Talat Masood, 'The mobility and even the survivability of the land forces depended to a large extent on air cover and so does the security of the naval forces. Superior air power is thus vital for the success of any military engagement.'[4] Air Force officials wanted to make up for its inferiority through procuring superior technology. In the absence of American sources the negotiations were held with the French for the procurement of the Mirage 2000-5. In the opinion of the Deputy Chief of Air Staff (Operations), the connection between the American and the French aircraft for the Air Force presents a game of 'see-saw'.[5] Whenever the service was deprived of American hardware it turned towards French equipment. The first time this method was adopted was in the 1970s, when the US arms embargo led to the PAF's purchase of the French aircraft.

The Mirage 2000-5 with its BVR capability, superior avionics, and armament was considered a good option to meet the threat. The Air Force was hopeful of acquiring these aircraft until 1997 when the President, Farooq Laghari, formally ordered the government to abandon a prospective deal on grounds of cost. The package would cost Pakistan more than $5 billion: this price was not affordable. There were other reasons too for terminating the deal. A particular assertion was that these aircraft were not meant for a ground attack role and could not use the NATO weapons that the PAF amassed during the 1980s. Moreover, it was claimed that an Air Force team of technical experts had rejected the aircraft several times because any technical alteration demanded by the PAF and promised by the manufacturer was not possible before three years after the delivery.[6] The Army viewed ground attack capability as vital in providing it with close-battle-support in time of war – the way it was done in the 1965 war. This perception was different from that of the Air Force, which was more interested in air superiority missions. A more serious dimension of the deal related to corruption charges. From 1990–91 to 1996–97 three successive governments were accused of negotiating kickbacks with the French. Two Defence Ministers, Ghous Ali Shah and Hazar Khan Bijrani, were accused of personal involvement, and so too was Benazir Bhutto during her second term as Premier. Reports published in 1993 accused the Chief of the Air Force for pushing a decision on the purchase of the Mirage 2000-5 on personal grounds. One important member of the evaluation team, as it was asserted, was related to the air chief and was also one of the top executives of the Mirage manufacturing concern in Paris.[7] Such allegations, however, do not appear to be the main cause of failure in finalizing a deal. Any prospective deal was not viewed very kindly by the largest service – the Army – which had been deprived of resources to make quality weapons acquisitions for a long time.

A greater problem was the lack of resources to finance the purchase. Islamabad also tried to persuade Washington to return the $658 million that had been paid to procure the F-16s. By 1991 17 manufactured F-16s were lying in the US. The Pakistani government insisted that if the US government was unable to give the aircraft it should return the money. Congress's acceptance of the Brown Amendment in September 1995 paved the way for Washington to return the money by selling the 17 aircraft. But, even by 1997, the funds had not been returned because the US government had failed to re-sell these now *old* aircraft. Moreover, with improved India–US relations, Washington did not want Pakistan to acquire a technology that would disturb the regional military balance. It was, therefore, decided to supply Islamabad with American wheat instead. This made it impossible for Islamabad to buy the French aircraft that were drawing a lot of criticism at home. People started to question the need for acquiring such expensive technology. Such opinion did not take into consideration the technological advancements or the Air Force's urgent requirement to get quality aircraft. Undoubtedly, the PAF did require a multi-role air superiority fighter, but the lack of transparency that accompanied the negotiations from the beginning was instrumental in jeopardizing the talks.

More vital was the fact that the Air Force high command was so focused on the French aircraft that they totally ignored other sources of procurement. For instance, the PAF's talks with the Russians for the Su-27 'Flanker' were stalled because of the Mirage 2000-5 negotiations. People from the Foreign Office were of the view that Islamabad failed to acquire these aircraft owing to the well-entrenched pro-Indian lobby in Moscow.[8] The Vice-Chief of Air Staff did not comment on this. He, none the less, expressed interest in procuring the Sukhoi aircraft of the Su-27 series.[9] The fact is that the PAF got serious about the Russian aircraft very late. The main interests were the French and American aircraft, and a serious attempt to obtain Russian technology was not made until 1997 when it became clear that it was not possible to procure Mirage 2000-5s. This reluctance in the initial years of the 1990s resulted in creating an impression on the Russians that the Pakistan government was not a serious client, thereby weakening the relatively pro-Pakistan lobby in Moscow that was keen to sell to Islamabad. By 1999, the situation had become even more complicated. An increase in tension between the two countries over the outstanding issue of Russian prisoners of war and Islamabad's support of the *Talibaan* angered Moscow. While the Foreign Office[10] confessed that such problems hampered the smoothing of Pakistan–Russia ties, the military seemed oblivious of these realities. In their opinion, Moscow was too eager to sell weapons.[11] The only problem they found was Indian influence but felt that too could be circumvented through obtaining Russian technology from China. It was believed that once China got the technology to assemble the Su-30 aircraft, it had acquired from Moscow, it would not be difficult for the PLA to pass it on to the PAF.[12] After 1993 efforts were also made to procure the Swedish Grippen but talks did not materialize because the aircraft was fitted with an American engine and Washington did not allow Sweden to transfer these aircraft. Americans were of the view that this transfer would be tantamount to supplying F-16s to Pakistan. In addition, one American apprehension was that the sale of Grippens would annoy the US arms manufacturers who would criticize the decision to allow the transfer of Swedish aircraft using American engines when the local industry was not permitted to sell the F-16s.

Meanwhile, the PAF decided to upgrade its Chinese and French aircraft. From 1990–91 to 1996–97, three separate contracts were signed for the upgrading of the Chinese F-7s and Mirage III and Vs. The first was the contract given to the French manufacturer Sagem in 1990–91 commissioning the company to fit 'Integrated Navigation Units' in Pakistani Mirage aircraft. The second deal was to overhaul and upgrade the Mirage IIIOs that had been acquired from the Australian Air Force. This contract worth $118 million was signed in 1993. Again in 1994 the PAC, Kamra was given $116 million to overhaul 30 Mirage aircraft. This investment was in addition to $1.5 million spent initially on the Mirage IIIOs. The basic idea was to keep a certain number of aircraft functional so that the parity with the adversary did not diminish any

further. As part of this strategy, other sources were approached to negotiate the transfer of old/refurbished Mirage aircraft. Most prominent in
this regard was Islamabad's deal with the French manufacturer Sagem
to provide additional 40 Mirage IIIs. The first batch of these refurbished
and upgraded aircraft was delivered to the PAF in November 1997.
However, the last eight fighters were stopped after France imposed an
embargo in 1998. Another order was also placed for the Italian 'Griffo-
7' radar to be fitted in the cone of the F-7s. The Italian manufacturer
FIAR specially designed the radar for the Chinese aircraft having a full
look-up and look-down air-to-air capability through the use of Pulse
Doppler and medium PRF wave form, plus air-to-ground ranging
mode. Islamabad paid FIAR $20 million for manufacturing the new
system. It was later that a contract was signed for the transfer of technology for this radar to be manufactured by AWC.

Also in the 1990s a study was initiated for an overall upgrade of the
F-7s. The project called 'Super-7' involved the installation of Western,
mainly West European, avionics in the aircraft and an increase in the
payload. A report published in one of the leading Urdu language newspapers claimed that this target had been achieved, and that Pakistan
and China had managed to develop a new aircraft called the 'F-C1'.[13]
In talking to Air Force officials, it was discovered that the news was no
more than propaganda, and the air headquarters was still struggling
with the upgrading of the F-7 into 1999.

A project was launched in 1989-90 for the co-development and co-
production of a jet trainer with the Chinese manufacturer CATIC called
K-8. It was publicized as an indicator of growth in Pakistan's aircraft
manufacturing capability.[14] Such official statements were far from the
truth and did not reveal the problems encountered by the PAF in this
alleged co-development and co-production deal. The project was
marked with an absolute lack of transparency at all levels. In 1997 the
then Deputy Chief of Air Staff (Operations) said that the Plans and
Operations branch at the air headquarters was investigating the deal
since it was not clear why the previous management had signed a contract that did not benefit Pakistan. There were various gaps in the deal.
The transfer of technology element was minimal and there was disagreement between the collaborators regarding the choice of engine to
be used in the K-8. The Pakistani side wanted an American 'Garrett'
engine, an idea not welcomed by Beijing[15] which preferred a Russian
engine instead. Secondly, no evidence was found to support the claim
that Pakistan would co-develop the aircraft. Overall, there were no signs
of a dynamic working relationship between Islamabad and Beijing and
it could, at best, be treated as a case of a 'latent' collaboration.

The literature on collaboration shows that a dynamic relationship requires an element of complementarity, perception of balanced benefits from the alliance, regular development of responsibilities between the partners, adoption of a philosophy of constant learning by the partners, complementary assets, existence of synergies between the partners, approximate balance in size and strength, and compatible cultures.[16] An example of a dynamic defense industrial collaboration is the US–Japan joint project to co-develop and co-produce FS-X fighter aircraft. Both countries agreed to share some of their technical know-how regarding the development. The US partner, General Dynamics, had agreed to collaborate because the Japanese manufacturer had the know-how of developing a large wing box in a single piece by using the co-curing process – a technology in which Japan was more advanced than its partner. What facilitated this alliance was Japan's past experience of producing jet aircraft. This kind of technological parity and experience could not be found in the case of the Pakistan–China collaboration for the K-8. Pakistan had no experience of manufacturing jet aircraft that could benefit the project or get dividend out of a defence industrial collaboration.

In the case of the K-8, there was an imbalance between manufacturing capabilities, investment and responsibilities of the partners. Unlike China, which has a fairly long experience of manufacturing fighter jets, Pakistan had no skills in this area. Pakistan's main contribution was in terms of the PAF's operational experience. The Air Force staff worked closely with Chinese designers to work out a design for the K-8 providing 14 aeronautical engineers to work on the K-8 along with 2000 Chinese engineers.[17] This collaboration did not make up for the technological backwardness. In 1999 the design has yet not been frozen. The Chinese manufacturer completely controlled the designing and production with no sensitive technology transferred to Islamabad, who had only a 25 per cent share in the airframe. At various times the government tried to negotiate an increase of share to 45 per cent but Beijing was not responsive to its partner's demands. The collaboration was not based on any offsets. PAC, Kamra was only to manufacture 467 components of the airframe for aircraft procured by the PAF. All electronics, landing gear, and other components would come from China. Beijing was only interested in the project because it would give China a chance to increase arms sales. In 1996–97 Islamabad paid $20 million for the transfer of six prototypes K-8. An additional benefit was that such transfers, categorized as co-development projects, would not count as sales, thus were not recorded in the UN arms trade register. This benefit was for Beijing and not for Islamabad.

What made the K-8 deal more interesting were the varied views of the PAF officers about the project. One responsible source told the author that the Chinese started the talks regarding K-8 in the 1980s but PAF officials did not show any interest, the then air chief resisting the deal despite pressure from the MoD. The officials in the Ministry were of the view that this collaboration was vital for Pakistan's defence industry and it insisted upon the Air Force showing its commitment by buying at least a few K-8s, still under development.[18] Air Chief Marshal Hakimullah had resisted because, technologically speaking, the K-8 was of a lesser quality than the American T-37. In addition, he was not willing to buy the prototypes. His decision was reversed by the next air chief, Farooq Feroz Khan, who agreed to procure six K-8s in 1994. The Assistant Chief of Air Staff said that the decision was peculiar because the PAF was not in need of a jet trainer.[19] The Director General Inter-Services Public Relations (ISPR), the official spokesman for the military, did not agree with this opinion being of the view that the Air Force needed a trainer aircraft since there were constraints in using American trainers owing to the embargo.[20] The PAF was operating American T-37s, which were procured from the US on lease. Over the years, about 34 were purchased while 19 continued to be on lease. Considering Pakistan's resource constraints it was considered a viable option to co-produce an aircraft with Chinese help which would have the additional benefit of saving funds for investing in R&D for developing a totally indigenous aircraft.

Another project started by the Air Force was the creation of the Air Weapon Complex (AWC) in 1992–93. The idea was to establish an organization that would operate differently from a traditional public sector organization. Unlike other facilities, AWC was given a free hand in using its budget. The management aimed at developing core technologies that would help in manufacturing a variety of products it could market through an independent marketing wing. The Director of the Complex, a serving Air Force officer of air commodore rank, later promoted to Air Vice Marshal, claimed that by adopting this strategy, the facility was able to make inroads into the commercial market.[21] In early 1999 an Unmanned Air Vehicle (UAV) was exhibited at the first naval defense show at Karachi. This equipment, it was claimed, was much cheaper than those made by other countries. The fact is that the AWC imported a lot of technologies and components from South Africa that they later presented as indigenous development. The UAV itself was not on a par with similar equipment produced by other countries. It was an under-developed product with the capacity to carry a weight of no more then 30 kg. Moreover, a large number of components were to be imported from foreign sources like South Africa.

Weapons procurement for the Army

This period was equally frustrating for the Army. Despite being the largest service and enjoying a prominent position in the country's power politics, the service did not manage to acquire any worthwhile major weapon system until 1996–97.

By 1990 most of the tanks in the Army's inventory were qualitatively inferior Chinese tanks. The American tanks M-48A5s were refurbished but old. The Army thought of redressing this problem by acquiring more Chinese tanks, overhauling and rebuilding them locally, and later manufacturing an indigenous main battle tank. General Mirza Aslam Baig, who took over as the Army chief after Zia's death, had ambitious defense production plans. He had encouraged a plan whereby Pakistan would focus on developing an MBT with Chinese help and, in the meantime, acquire foreign-made tanks to fill the gap. As part of this strategy Chinese T-85s were procured in 1990. This tank, it was believed, could also provide a technological cushion between the less sophisticated T-59s, T-69s, T-69IIs and the more sophisticated MBT-2000. The Chief of General Staff stated the number of the tanks at 70–80,[22] while another source[23] quoted a figure of 300. The discrepancy in figures could possibly be related to the negotiated figure and the actual acquisition at the time of the interview. The acquisition began in 1993.

Meanwhile a contract was signed with the Chinese manufacturer Norinco for the co-development and co-production of the MBT-2000. In making such ambitious plans, the Army top brass, represented by General Mirza Aslam Beg, obviously did not think about the industrial and technological discrepancies that hindered Pakistan's defense production. 'Al-Khalid' did not meet its production target. The second and third prototypes were rolled out before the mid-1990s with the fourth one scheduled to be tested in the summer of 1998. Beijing and Islamabad could not decide on the power pack to be used in the tank, but in 1997 the Pakistan Army finally ordered the Ukrainian 1200 engine that was fitted in the T-80UD tanks. According to the Director General HIT, the Army was pleased with the performance and cost of the Ukrainian engine worth $0.25 million per unit against the initial choice of the American Perkins engine at $1 million per unit.[24]

While the development of the MBT-2000 was taking time, the decision was made to procure foreign tanks to fill the gap. Various sources were approached for the purpose, the most significant being Poland followed by Ukraine and Russia. Initially, a deal was negotiated with Poland in 1992 for 320 T-72s. The cost was approximately $450 million and the deal was comprised mainly of barter arrangements. Islamabad was expected to pay 10 per cent of the total amount as down payment, and the total payment was to be completed in eight years through barter trade. The deal was cancelled at the final stages, long after the tanks had been approved in the trials held in Pakistan. The reversal of this decision coincided with the sudden death of the Army chief, General Asif Nawaz Janjua in 1992. Analysts were of the view that this happened because General Janjua was the main, and perhaps the only, supporter of the decision having endorsed it because of personal interests.[25] According to Mushahid Hussain, this decision was directly linked to the commercial motivation of some people in the Army's policy-making elite. Although there is no hard evidence available to prove such an allegation, one can draw certain inferences about the decision-making process and the role of key actors in influencing decisions. The fact that, despite the Army's need for tanks, a deal was cancelled in its final stage, soon after the death of the Army chief, reflects the influence of this office-bearer on procurement decisions. This also proves that arms procurement policy-making is less institutionalized than one would imagine. Soon after another deal was negotiated for the Ukrainian T-80UD tanks that were superior to the Polish T-72s. Finally, in 1997, a contract was signed for 320 tanks, obtained for $1.6 million per unit thus accumulating a total cost of $600 million, the transfer creating the possibility of another deal for the tank engines. The total cost also included the cost of training and spares. The deal was not completely 'clean'. The fact was the Ukrainians had cannibalized about 80 of the older tanks to manufacture 30 units for Pakistan but later encountered problems in fulfilling the contract. In 1999 the contract was still under completion. This indeed strengthened the rumours of the kickbacks in the deal. Some of the components of these T-80UD tanks were being manufactured in Russia. Apparently the Ukrainians had to aggressively negotiate with Moscow for the release of these components to complete the Pakistani order. The question that arises here is why did the Pakistani team not take account of the Ukrainian limitations? Did Ukraine appease the team by offering them kickbacks?

In 1990 a deal was signed with the American manufacturer FMC for the transfer of 775 kits of the M113A2 APCs to be assembled at HIT. This was the only piece of equipment to be transferred to Pakistan after the arms embargo because the manufacturer had removed the machine gun mounted on the APC, which came under the sanctions. With this action the APCs no longer fell into the category of weapon systems. The HIT staff claimed that their facility had started producing APCs, but on further investigation it was found by the author that the M113P, as the APCs were renamed, were those assembled from the 775 CKDs procured from the US, making HIT's claim entirely spurions. The facility's only contribution was in the form of adding external fuel tanks and fitting it with a Chinese machine gun. In 1997 the decision was taken to import another 1000 kits from the US.

In 1991–92, during Prime Minister Sharif's first term in office, a proposal of closing down Margalla Electronics was considered. The radar production facility, as was pointed out in Chapter 6, was established without any clear justification. By 1991–92 there wasn't much happening at the facility but a decision to close down the facility, none the less, was averted by acquiring two 'Giraffe' fire control radar in CKD form in 1993. The Army and MoD officials could only sustain the establishment if additional work was generated. To sustain Margalla Electronics, negotiations were also started with China. Nothing is known about the outcome of the talks, Beijing being more interested in selling Islamabad CBUs, SKDs, CKDs, and strategic spares and components. For example, China supplied M-11s to Pakistan that were renamed 'Hatf-III'. These missiles came with a shelf-life until year 2000. The Chinese were not forthcoming in providing its ally with the technological know-how to extend the life of these missiles.

The most significant feature of the period from 1990–99 was Islamabad's frustration over its inability to fulfill the requirements of its defense forces. The military top brass and the political leadership were perplexed at the US arms embargo. Washington's policies were construed as an act of letting down Pakistan at a time when Islamabad's adversary, India, strengthened its military power both quantitatively and qualitatively. What worsened the situation was Pakistan's resource constraints. It did not have sufficient funds to buy major weapon systems and, needless to say, whatever funds were available were wasted in acquiring systems and technologies that did not strictly qualify as 'force multipliers'.

9
Mutually Assured Deterrence: the Nuclear Option

Islamabad had started to work on developing its nuclear capabilities in the 1970s, the program gaining speed after India's nuclear explosion in 1974. The strategic imperative was paramount in establishing the nuclear program. With time, nuclear deterrence gained importance in the military strategic planning leading to a juncture where non-conventional defense was considered as the only viable option to fill the gaps in the country's conventional defense.

Besides its links with conventional arms capability, the nuclear project also tells a lot about the dynamics of defense decision-making in Pakistan.

Pakistan's nuclear option: the strategic dimension

Islamabad developed its nuclear capabilities as a reaction to India's explosion. Dougherty aptly explained this reactive strategy. 'Proliferation by reaction is a phenomena [sic]', he viewed, 'associated with pairs of conflict-parties or historic rivals rather than a chain-reaction involving indefinitely long series of countries. In "proliferation-by-reaction model", if one country acquires [the nuclear weapons], the traditional foe feels itself under compulsion to acquire [the nuclear weapons] for the sake of protective equilibrium.'[1] A latent capability could prove an effective safeguard against Indian nuclear blackmail to deter New Delhi from aggression and neutralize its regional dominance. This faith has gradually increased. A popular notion in the policy-making circle was that only the possession of a nuclear capability had minimized the possibility of a future war in South Asia.

Considering Pakistan's financial constraints and its inability to narrow the conventional military technological gap against India, the top leadership, including the civilian and military, developed greater faith in the nuclear option. This reliance increased especially after 1990 when the government could not carry out major weapons modernization because of political and economic constraints. This view was shared both by military personnel and the political leadership who wanted to use this advantage when it launched a low-intensity offensive in Kargil in the summer of 1999. The plan was to acquire limited territory at Kargil and gain victory in a tactical maneuver forcing India to confront the issue of solving the Kashmir problem on terms favorable to Pakistan. More important, Pakistan's nuclear option would limit India's choice of launching a bigger offensive the way it had done in 1965. Interestingly, with every injection of technology Islamabad has tried to solve the Kashmir issue militarily. In certain ways the Kargil offensive was reminiscent of 1965 when American arms transfer had given the armed forces ample confidence to launch a limited offensive in the Runn-of-Kutch area to resolve an outstanding issue. The fundamental difference between 1965 and 1999 was that now, unlike 1965, India would have to be careful crossing the international boundary, A development attributable to Pakistan's nuclear capability.

South Asian nuclear deterrence was based on a peculiar cost–benefit analysis. Both India and Pakistan worked out a rather simplistic equation which can be explained by analysing Dr A.Q. Khan's comments given in his 1984 interview with an Indian journalist, Kuldip Nayyar. This was the first time any Pakistani source had spoken out about certain details of the country's nuclear weapons capability that had created a sensation on both sides of the border. The interview was significant because it presented Pakistan's 'nuclear doctrine', which was that while Pakistan would need about five nuclear devices to target an equal number of Indian cities, India required three to four bombs for a similar action. According to this strategy, Islamabad could threaten important and populous Indian cities such as Delhi and Bombay, which have a combined population of 22.75 million (1991 census). India could do the same to at least three Pakistani cities – Rawalpindi/Islamabad, Lahore and Karachi with a total population of 20.1 million (1998 census). It was assumed that potential destruction of this magnitude would dissuade both countries from a nuclear encounter or any conflict that could lead to a total war. This standpoint was subscribed to by other people such as the former Foreign Secretary, Niaz A. Naik, who also played an important role in track-II diplomacy with New Delhi on behalf of Islamabad during the crisis period in 1999. In discussion with the author, Naik conceded that the idea was to have at least one nuclear device that could be delivered on at least one of the highly populated Indian cities. This threat would suffice in deterring the Indian government from waging a war on Pakistan.[2] Naik's limited approach also depicts the greatest limitation of the nuclear option: the absence of a nuclear doctrine. Even India's nuclear doctrine presented in 1999 was incomplete. Although India used terms like 'minimum deterrence' and 'no-first use of nuclear weapons' as the guiding principle behind its nuclear doctrine for which it had planned to spend about Indian Rs. 700 billion, the policy was open-ended. It did not specify exactly the threat that the Indian government must counter with its nuclear capability. Information on weaponization, threat specification, and plans for a command and control system to operate the nuclear weapons, if the need arose, was not provided. Pakistan's nuclear planning was even more rudimentary than India's. While stating that nuclear weapons would only be used in defense of national security, there was no clear indication of when Islamabad would use these weapons. Ballistic missiles such as 'Ghauri-II' with a range of about 3000 km was under development. Whether Pakistan would need to develop an ICBM capability, is a question that cannot be answered unless Islamabad produced a nuclear doctrine. The main focus of Pakistan's nuclear plan is India and to force New Delhi to give up Kashmir. One notion was that Pakistan had always had a nuclear doctrine: in the 1980s Pakistan had used its nuclear option to manipulate the US into providing Pakistan with conventional weapons. It was during the same period that India was twice deterred from any aggressive action because of the threat of a nuclear response from the adversary.

Such actions were viewed as presenting Islamabad's nuclear doctrine. This approach, however, was political in nature and did not take into account the technological necessities for developing a credible nuclear deterrence. For instance, despite announcing a nuclear role for its Navy, Islamabad did not venture to carry out tests to modify the warheads to be used from the naval platforms. By the end of 1999, there were some plans to deploy static missiles on naval ground installations, but would thus actually ensure a second strike capability? More important, are there any plans at all for developing a second strike capability? If that is not the case then what kind of deterrence do military planners have in mind? After going overt with the nuclear capability it is imperative that New Delhi is convinced of Islamabad's ability both to protect itself and strike back. Moreover, India and Pakistan did not have the technological wherewithal such as the permissive action links or second strike capability to make nuclear deterrence credible. It was believed that Islamabad could convince the international community, especially the US, of the need to provide it with the technology. In a conference on disarmament held in Albuquerque, New Mexico in April 1999, Pakistan's ambassador to the UN, Ahmed Kamal, stated that if the US government was concerned about safety of nuclear weapons in South Asia it should provide Pakistan with technology like the permissive action links.

Technologies were developed for which direct US support was not required. Delivery systems was one such area. In 1990 trials were carried out to deliver a nuclear device from an aircraft. Reportedly, it was one of these simulations that was photographed by an American satellite and construed as Islamabad's preparedness to launch a nuclear attack.[3] In 1998–99 two IRBMs, 'Shaheen' and 'Ghauri', were developed with the potential to carry nuclear warheads.

The primary feature of South Asian nuclear deterrence is ambiguity. The basic idea is to impress the adversary by consciously but carefully publicizing the country's nuclear capabilities without disclosing any information about the actual strength. A party involved in a conflict when unsure of its adversary's capabilities, and considering its own capabilities, might decide to launch a first strike. In such a situation 'first strike' could result from 'nervousness' rather than 'calculation'[4] and could be extremely threatening for neighboring countries. A country encountering a first strike would be left with no warning time prior to an attack.[5] Thomas countered this argument by stating that geographical proximity, plus the interrelated nature of societies on either side, as in the case of India and Pakistan, had always helped in deterring their policy-makers from carrying out such extreme action.[6] His argument was based on the assumption that policy-makers always make decisions logically and dispassionately, without any chance of miscalculation. In view of the tensions in the summer of 1999, Thomas's notion was indeed questionable. What if New Delhi decided to cross the international boundary between India and Pakistan in the wake of the Pakistan Army's action in Kargil? What if the US had not put pressure on Islamabad to withdraw its forces from Kargil? In a discussion with the author the former Indian Naval Chief, L. Ramdas, said that the Indian Navy was considering a blockade of Karachi if the situation had not diffused in Kargil.[7] How would Pakistan have reacted? One could run through a number of scenarios – all indicating a potential military conflict and a nuclear disaster. In the eyes of policy-makers, nuclear capability made Pakistan appear stronger and more muscular. These attributes are deemed essential in not only resolving outstanding issues with India but also in being taken more seriously by the international community, which was not inclined to understand Islamabad's standpoint, especially on Kashmir.

The crisis in Kargil in late 1999 was indicative of Pakistan's nuclear doctrine. The understanding in Islamabad was that the time had come to use the nuclear umbrella to force the resolution of the Kashmir issue without risking a conventional war. It was thought that, given Pakistan's nuclear capability, India would not dare respond militarily to Pakistan's Army's tactical maneuver. What was not taken into account were India's diplomatic capability to thwart such designs and the international community's pressure in not allowing conflict escalation in South Asia. Therefore, the Clinton administration forced Pakistan to withdraw its troops from Kargil.

It was also linked with the fear of India's winning permanent membership of the UN Security Council on the basis of its nuclear capability – a reason for Islamabad's immediate reaction to India's nuclear tests in 1998. Pakistan did not want its adversary to win an elite status. Although American analysts and policy-makers ruled out the possibility of India's entry into the Security Council,[8] the growing significance of New Delhi in Washington's strategic and political assessment was not reassuring for the Pakistani leadership. Their worry was that Washington would not hesitate to include India in the elite club of the international community.

In one respect, development of Pakistan's nuclear capability was linked with its relations with the US. American policy on nuclear proliferation was inconsistent particularly regarding South Asia. In the early 1970s, Pakistan had warned the US of its fears of proliferation at the time of India's PNE but nothing much was done. Islamabad, which always wanted a multilateral arrangement for solving its bilateral problems with India, was discouraged by Washington's stance on the issue. American criticism of Pakistan's nuclear activities was seen as an act of singling out Pakistan. Islamabad–Washington relations had also worsened towards the end of the 1970s owing to their diverse views on nuclear proliferation and Pakistan remained conscious of the fragile Pakistan–US security linkage during the 1980s as well. Islamabad, therefore, could not trust America to help in attaining a comfortable military balance with India. In terms of time, the strategic situation of the 1980s had provided Pakistan with a gap that could be used not only to enhance conventional capabilities but also to accelerate work on the nuclear program. The policy-makers were sure that Washington's act of ignoring Islamabad's nuclear activities depended on the Soviet presence in Afghanistan. They did not know at that time, nor did the Americans, that the USSR would collapse within the decade. Towards the 1990s this fear not only seemed to deepen but it was also coupled with the anxiety that India might acquire more significance in the broader American-power-framework for the region. This perception strengthens the belief that it was strategically viable for Pakistan to have a nuclear option especially in a situation where it could not depend on America *vis-à-vis* India, but it was also seen as one way of keeping Washington interested in Pakistan and South Asian regional geo-politics. The ambiguity factor was maintained despite both neighbors going overt with their respective nuclear options in the summer of 1998. Both New Delhi and Islamabad were unable to announce a solid nuclear doctrine.

From the 1980s, the nuclear activity became increasingly linked with the Kashmir issue. Islamabad encouraged the impression that the fate of nuclear proliferation or a nuclear war in the region is related to the settlement of the Kashmir issue. By doing this, it was hoped that countries of the world, particularly the US, would put pressure on India to solve the outstanding problem. From a strategic standpoint, the India factor was vital, especially in the 1980s, in order to divert international pressure from Islamabad to New Delhi.[9] Of course, Indian gestures such as the refusal to sign the CTBT helped Islamabad's stance. New Delhi was averse to the idea of linking the Kashmir problem with nuclear proliferation, fearing that this would invoke more international pressure on India to roll back or cap its nuclear agenda.[10] This in turn forced New Delhi to come to terms with solving the Kashmir problem, which was exactly what Islamabad had tried to achieve in 1999 through its military maneuver in Kargil. The military operation failed to achieve the core objective but it did support Pakistan's assertion that its nuclear capability could deter India from waging a full-scale war the way it had done in 1965 in response to a similar military operation by Pakistan.

Nuclear proliferation: the domestic political perspective

It would be unfair not to mention the personal ambitions and organizational imperative that also drove the nuclear program. Zulfiqar Ali Bhutto's ambition to become a leader of international stature played a detrimental role. Bhutto wanted to become the leader of the Muslim and Third World through projecting Pakistan's military prowess. A nuclear weapons program, in his view, could turn Pakistan into a world power,[11] thereby making him an important world leader. A mix of personal and political reasons continued to motivate the country's leadership in the 1980s and 1990s as well. It was Zia's drive to gain popularity within the country that influenced his continued support for the program. The General was afraid of the deposed political leader's popularity, and the possibility of Bhutto's returning to power. His military regime needed legitimacy and popularity, which he hoped to acquire by supporting the popular, albeit controversial, agenda of nuclear proliferation for enhancing national security. One can certainly not ignore the strategic significance of the nuclear program for Zia's government. The political involvement of the military in the 1980s made it even more vital to keep defenses strong and nuclear proliferation was the only feasible course of action.

General Zia's use of the country's nuclear ambitions for political purposes was to set a trend for future regimes too. Prime Ministers Benazir Bhutto, Nawaz Sharif, Sher Baaz Mazari and Moen Qureshi, despite their varied ideological backgrounds and inclinations, also supported the program. Benazir Bhutto had the most controversial stance regarding the nuclear program as she was known to have opposed the Army's plans regarding the use of the nuclear option, but this was during her first stint as Prime Minister. In her second term, she forcefully adhered to the military's perspective. She refused an American request to verify the capping of Islamabad's nuclear capability. It was quite clear that she had realized that capping, rolling back or abandoning the program would be tantamount to a political suicide, a policy understood by other regimes as well. This decision-making pattern can be better understood using Meyer's analysis: national decision-makers can use nuclear proliferation to direct domestic energies away from domestic problems, and to enhance domestic morale in the face of civil strife, ethnic hostility, and so on.[12] This led even Nawaz Sharif's government, with a two-third majority, to support the nuclear option.

No political leader, however, was willing to evaluate the cost of nuclear deterrence, especially the opportunity cost for socioeconomic and human development needs of the nation. Despite the millions of poor and malnourished people, Islamabad was not willing to give preference to social and economic securities over military security. Building a nuclear program at the cost of socioeconomic development could technically be afforded in a country like Pakistan, where about 80 per cent of the population was dependent upon homegrown agricultural produce. With little exposure to a higher standard of living, the majority of people was easily swayed by high-sounding political slogans and emotional rhetoric carried out in support of the nation's nuclear capability. It was convenient for Prime Minister Nawaz Sharif to impress the populace by claiming that his decision to conduct the tests was to safeguard the country against a superior external threat. A similar situation was found in India's case as well where the common populace was heartened by images of the country's military strength.[13] This situation in the Indian subcontinent is likely to continue since India and Pakistan are self-sufficient in agricultural produce and basic eatables, and it would take a long time before the deprived people realized the cost of nuclear deterrence. This factor also played a major role in policy-makers' not feeling any pressure from the public to review the national security agendas.

There was a certain consensus amongst the populace and the leadership to support nuclear proliferation at the cost of other needs. There were hawks like Begum Abida Hussain, an important member in both Nawaz Sharif governments, who believed that Islamabad must retain its nuclear option for attaining long-term economic growth goals. In her view, territorial security was a prerequisite for economic and social security which, in Pakistan's case, could not be attained without nuclear deterrence. The non-conventional defense capability would provide Islamabad a strategic respite against India, allowing Pakistan the time and opportunity to strengthen its economy.[14] In presenting these views she totally ignored the collapse of the Soviet Union as a result of economic pressures. The leadership was not moved by these lessons and the list of leaders who advocated a nuclear option for political reasons was exhaustive. Their primary concern was to appease the Army, which was not willing to give up the nuclear option. After the Kargil crisis, the Army chief tried to present the political government as being solely in charge of nuclear decision-making but the reality was otherwise. In a statement made a little before the coup, he stated that it was the government's decision whether to sign the CTBT but such claims were to increase complications for the Sharif regime. The issue had serious emotional and psychological bearings and, by giving such public statements, the Army chief aimed at restraining the political government from compromising nuclear ambitions. The emotional angle of the nuclear debate is paramount in both India and Pakistan, and this factor alone makes it difficult to carry out a debate objectively.

Nuclear proliferation: the bureaucratic perspective

The military's control of the nuclear program is one of the main explanations for Islamabad's consistent support of its nuclear option. After

1977, the country witnessed political turbulence as nine regimes including two caretaker governments were changed. Under these circumstances, the Army was the only institution that could provide continuity. It was, therefore, obvious for the Army to support non-conventional defense more than any other individual or group in the country. Although the Army's inclusion in the project had been in evidence from the beginning, it increased after Zia took over in 1977. The financial and administrative issues, however, had been controlled by the uniform personnel. This was in addition to the engineering services provided by them. In many respects the Army personnel look at the nuclear program as an extraordinary endeavor that could not be undertaken without their commitment. The Army's management, hence, is skeptical of any political leadership's control of the program. It is feared that politicians, who may be less committed than the Army to sustain a nuclear capability, would compromise national security by adhering to the non-proliferation agenda of the international community.

As mentioned earlier the Army gradually established control over the program. Bhutto started the nuclear project in the early 1970s, at a time when the Army's involvement was limited to technical support. A team working at the POFs, Wah under the code name 'Research' later known as the 'Wah Group' had responsibility to carry out R&D on explosives to be used in the nuclear device. This small group of nuclear scientists and engineers handled nuclear technological know-how *in toto*, its control continuing until organizational politics began to play a role within the nuclear bureaucracy. For instance, Dr Khan, who had managed to carve a place in the nuclear program by bringing much needed information on uranium reprocessing, succeeded in lowering the esteem of other members of the PAEC including its Chairman, Munir Ahmed Khan, who was reprimanded several times by Bhutto for not meeting deadlines. For Bhutto, it was important to obtain the nuclear capability at all costs and in the shortest time possible. He, of course, did not appreciate the fact that it was not an easy task to develop this technology.

Dr Khan's role in making the nuclear project take off is unequivocal. It was because of him that the choice was made to follow the uranium enrichment path rather than go the plutonium route. He was duly rewarded by successive regimes. Although a metallurgist, he earned for himself a reputation for being a nuclear scientist. He managed to smuggle in the know-how for uranium enrichment from Holland where he had been working at a URENCO uranium enrichment plant prior to returning to Pakistan. There, he managed to obtain sensitive centrifuge technology for uranium enrichment. Dr Khan worked on translating some German documents into Dutch on German centrifuge while he was in Almelo, Holland from 1972–75. Information from him provided a breakthrough for Pakistan. The centrifuges working at Kahuta, therefore, are the G-1 and G-2, an early-generation German technology. In addition, he had managed to obtain information on subcontractors supplying gas centrifuge equipment to URENCO, which was also useful in acquiring similar hardware. There is no evidence that Dr Khan was planted in Holland by the Pakistani government; nevertheless, it was during his stay in Holland that he became involved in Islamabad's plans and he offered his services. In a letter, written to Prime Minister Bhutto in 1974, he showed his willingness to help Pakistan attain the non-conventional defence capability. While providing this help he also managed to earn a place for himself in nuclear decision-making. The availability of such information was clearly the reason for Pakistan's adopting the comparatively difficult path of making a 'fission' weapon by enriching uranium rather than embarking upon plutonium reprocessing. In the words of Muhammad Aslam: 'Kahuta is solely A.Q. Khan's baby'.[15] He was, in fact, the only person involved with the program to have enjoyed the limelight. Other major contributors were resentful of the attention given to Dr Khan while they were left unacknowledged.

Dr Khan's competitor was the PAEC, especially the Commission's 'Special Development Works' group, designated in 1992–93 as the National Development Complex (NDC). The nuclear tests carried out in July 1998 were the work of the PAEC and NDC. Although responsible only for uranium enrichment, Dr Khan pretended, however, to be solely running the project. Over the years, Dr Khan managed to turn KRL into a self-contained bureaucracy with its independent R&D, production, and marketing wings with huge non-auditable financial resources at his disposal. The PAEC and KRL represented a duplication of activities caused by the confrontation posture of the two heads of these organizations. In fact, at the time of the nuclear tests Dr Khan wanted to take the lead by carrying out the tests and was quite annoyed when the job was given to the PAEC and its sister concern, the NDC.[16] A story carried through the grapevine in the capital was that the Prime Minister had first asked Dr Khan if he could respond to the Indian tests and the time required, to which Dr Khan replied giving a schedule of one month. The Prime Minister was in a rush to get underway and it was then that the PAEC offered to conduct the tests sooner. Six tests were undertaken within about fifteen days of

India's explosions. Interestingly, the head of NDC, Dr Samar Mubarikmand, hailed from the same province as Nawaz Sharif and this link provided the necessary ties that helped him develop a better rapport with the Prime Minister. Mubarikmand was also part of the group opposed to Dr Khan. In 1999 there were rumors of the Sharif government investigating the siphoning-off of funds provided for the manufacture of the 'Ghauri' missiles. Apparently, only 13 missiles were in the inventory while funds were obtained for 20. What role the group opposed to Dr Khan played in revealing this information to the government is subject to investigation.

The impact of the competition within the nuclear bureaucracy on the final decision to go overtly nuclear, or nuclear decision-making, is a matter for debate. Unlike the Indian nuclear bureaucracy reputed to be comparatively more assertive, the Pakistani counterpart has a lesser influence over policy-making. The Army plays this role instead. Whether the Prime Minister was solely responsible for the decision to conduct nuclear tests or whether he did it in partnership with the Army is a question that was debated for months after the tests. What, however, was not disputable was the Army's role as a senior partner in nuclear decision-making. Since the nuclear tests and the announcement of a Nuclear Command and Control Authority in early 2000, the Army has emerged as the key actor in nuclear decision-making. The fact was that all the players involved in initiating a conventional military operation, deployment of the nuclear weapons, and taking the final decision to go nuclear were put on the table under the same umbrella of the nuclear decision-making Authority. This would certainly make anyone nervous. Given the Pakistan Army's present mood to face the Kashmir issue on India, such a system can be deemed threatening for peace in the region. A better option is to separate decision-making from implementation.

The technical experts probably enjoyed more significance than the political leadership. This was primarily because, alien to technological issues, the Army chose to leave technical matters in the hands of the relevant experts. Army personnel could be heard murmuring about the independence of KRL in getting funds from the Army that were used without GHQ being able to exercise sufficient control over the usage of these resources. This issue, however, never became a bone of contention between KRL and the Army. The two segments, the Army and technical bureaucracy, enjoy a sound relationship because, while the Army was forthcoming in providing support and finances, the latter did not object to the administrative and financial control of the service. Dr Khan, Dr Mubarikmand and others enjoyed a degree of autonomy in running their organizations despite the presence of uniform personnel.

What allowed the technical experts to enjoy this independence was, however, their ability to develop nuclear deterrence that the Army saw as the only answer to narrow the widening military technological gap with India. The military managers' singular concentration on military security provided no other option but to follow New Delhi in nuclear proliferation. In case of a conventional war with its adversary, Islamabad had sufficient reserves to sustain for about a week and no more. The economic situation made it less likely that conventional deterrence would be strengthened. Non-conventional defence, hence, was viewed as a protective umbrella to stop any further Indian onslaught but this approach was not straightforward because nuclear weaponization itself requires commitment of resources that Islamabad might not easily find. Would the government then opt for a simplistic model of deterrence proposed by some of its policy-makers of building a few weapons to counter India's nuclear arsenal, and how effective would this option be, are questions that the policy-making elite needs to address. Another alternative would be to sign the international non-proliferation treaties but this would not be possible unless the leadership is willing to carry out a review of national security and redefine its security objectives. A reassessment of national security objectives would not be possible under a military leadership, requiring a more responsible and honest political government able to carry out a cost–benefit analysis security policy. This re-evaluation would not be based on a western strategic ethos but rather an assessment of the broader societal needs of the country.

10
Looking Ahead

Towards the turn of the twentieth century Pakistan's political, economic and strategic conditions remain unenviable. There was no change in threat assessment and commitment to military buildup continued. Military modernization, however, largely depends on financial strength and the ability to forge strategic alignments. This factor, in turn, is linked with the state of the political process in the country.

The need for military modernization

India remains Pakistan's main source of threat. Similarly, the primary form of conflict for which the military continues to prepare is a medium-intensity conflict to be fought on the land frontier. However, the role of conventional forces in Pakistan necessarily has been minimized to a defensive posture. Most of the equipment is old or refurbished and the last time that a significant military modernization exercise was carried out was in the 1980s. Although the PN procured new hardware in the 1990s, the Navy's acquisitions are of little significance for the existing military plans for the country. Military-strategic planning required first the Army and then the Air Force to have the technological capability to defend the national frontiers. The Army's capability to launch an offensive was reduced but it maintained the capacity to counter an Indian offensive. The short encounter with Indian forces at Kargil proved the effectiveness of the Army's Air Defense Command. It also proved that with the nuclear umbrella the conventional forces could keep India at bay for a limited duration.

The larger service had increasingly incorporated the nuclear capability in its operational plans. While the Army managed such an alignment of conventional and non-conventional technological capabilities, the other two services believe in building stronger conventional defense. The PAF, in particular, has continued to search for a more capable multi-role fighter aircraft. By the end of the 1990s the quality of the hardware of the service was indeed not up to the standards where the service would get an edge on its adversary. It was primarily dependent upon Chinese interceptor aircraft. The fleet of French Mirages was old and the operations of the F-16s were limited owing to unavailability of spares, some of the 32 F-16s being kept for special operations such as delivery of nuclear weapons. The effect of the embargo was obvious on the transport fleet as well, the PAF encountering problems in operating its C-130s. In the search for major weapons acquisitions, the Air Force tended to ignore other equally important areas such as electronic warfare equipment and smart bombs and ammunition. The service did not have the resources or the indigenous technological capability to acquire modern missile systems, its mainstay continuing to be American 'Sidewinder' missiles.

The Navy had a real problem with its old fleet of frigates. Erroneous decisions in the past had left the service with hardware that needed to be upgraded at a fairly high cost or else equipment that could not be used because of unavailability of spares. In June 1999, the government made an announcement regarding giving the Navy a nuclear role. This was in anticipation of the development of future capabilities. If India continues to develop a sea-based nuclear weapons capability, Pakistan may opt to respond in the same coin. In case of a nuclear technological proliferation, and given Pakistan's lack of strategic depth, setting up a sea-base nuclear capability is an option that Islamabad may seriously want to consider. Did the Navy plan to use its conventional Agosta 90-B submarines as delivery platforms? Although there was no evidence of such planning, this option would prove extremely risky. Important as it was to use the service for a second strike capability there were no obvious plans to integrate the Navy in the nation's nuclear planning. Did Islamabad want to continue with its dependence upon a static and purely land-based ballistic missile force? Since the 1980s, a number of ballistic missiles have been developed by the country, but this technology needs improvement in order to turn it into a credible deterrence. Increasing options would also be tantamount to the Army giving an equal status to the other two services in military planning. Developing a sea-based second strike capability requires procurement of Naval platforms capable of carrying nuclear warheads. This would be tantamount to diverting resources from Army to the Navy, a development that the larger service might not currently-support. In 1999 the two smaller services were asked to tender their views on the establishment of a Strategic Command. The idea of using the principle of seniority or on a rotational basis for the selection of the head of this organization was discarded by the larger service. In the end, the Army opted to re-designate its Combat Development Directorate into the Strategic Development Directorate. The military modernization agenda would be set primarily by the larger service and it would support only those decisions of the smaller services that assisted the Army in its plans. This would naturally lead to an increased gap in perception between the three services resulting in more financial wastage and rivalry. Any change in this situation is linked with the future of the Army's role in the country's power politics.

The political scene

The democratic process in the country proved extremely shaky. Events at the close of 1999 showed that power politics would continue to be dominated by the Army, the key player since the 1950s. The popularly elected government of Nawaz Sharif was overthrown in a bloodless coup in October 1999. Another popularly elected leader ousted by the Army was Zulfiqar Ali Bhutto way back in 1977. Interestingly, both leaders were known for their arbitrary policies, corrupt practices, and unfair dealing with opposition parties. Notwithstanding Sharif's policies, the coup was driven more by the Army's concern for saving its own institution. It was feared that by instituting personal control of the Army, Sharif would eventually downsize the military and reduce its influence. Moreover, the Army did not appreciate the Prime Minister's interference in military matters. Defense related issues have always fallen in the military's ambit.

Since 1985 the Army had influenced the removal of all four governments, twice dispatching the head of government because of interference in military affairs. The first victim was Benazir Bhutto, whose government was elected in 1988. She was sacked mainly as a result of her disagreement with the military on the selection of the Chairman JCSC. The new military ruler claimed his continued support for democracy, pledging in fact to remove fundamental lacunas in the democratic process. We know that Pakistan's democracy had not met its desired objectives owing to lack of accountability and political institutions but it is difficult to see how the military, which itself lacks accountability, would ensure institutionalization of such a process.

Despite the fact that many wanted to get rid of Sharif's arbitrary rule, the military action on 12 October came as a surprise. It was felt that the international political environment had changed to a degree that would not have encouraged the military leadership to contemplate direct rule. Pakistan's absolute financial dependence on the international community was indeed a crucial factor. The Army action showed disregard for the international political environment, especially opinion voiced by Washington. A popular belief was that the US manipulated political changes in Pakistan – a view attributable to Islamabad's intense interaction with America for over 30 years. The United States and Pakistan Armies represent two influences on the political culture of the nation, both believed to be capable of bringing about political changes in the country. Over the years, both the civilian and the military governments grew so dependent upon Washington's benevolence that any signal or statement needed to be carefully interpreted. However, by 1999 the communication gap between the US and GHQ at Rawalpindi had grown so wide that Army generals were less willing to pay heed to what Americans said. Pakistan's nuclear capability added to their confidence in operating independently of external influence.

Although the military has come under much criticism for interfering with the political process, in some respects it was forced to do so because of the irresponsible and immature attitude of the civilian leadership. The Pakistani military, which imposed upon itself the responsibility to defend the territory and ideology of the state, emerged as the main conduit of the establishment. It considers itself as the only entity that cares about the survival of the nation. Therefore, it was considered necessary and natural to interfere with political rule every time a civilian regime failed to deliver. The fundamental difference between military interference in the past and that in October 1999 was that the latest military coup was strongly motivated by organizational imperative. Sharif had adopted an erroneous strategy to curtail the powers of the Army through trying to breed dissension within the ranks. He tried to achieve this by promoting a very junior general, and one too who was from the engineering corps and not a fighting corps, to the position of Army chief. The Army's power could have been gradually decreased through strengthening certain institutional processes, such as, making the institution of the JCSC stronger through giving it budgetary control of the armed forces and firmly adopting the rule of appointing the senior most military officer as the head of this organization. The Prime Minister had violated this rule by not assigning the Naval chief, Admiral Fasih Bokhari, as the Chairman JCSC. Instead the Army chief, Pervaiz Musharaf, appointed one out of turn. The Admiral subsequently resigned in the fall of 1999 before the October coup. The political government had made the fatal mistake of not selecting a rare officer like Bokhari who did not share the military's contempt for civilian rule.

The previous Army chief, General Jahangir Karamat, was removed from office in 1998 as a punishment for broaching the idea of a National Security Council (NSC). This concept of the council espoused the approach of making the military a partner in decision-making. While providing constitutional cover to the military's involvement in power politics, it would also have laid certain responsibility on the armed forces. The idea presented by Karamat did not give the military the status and prestige accorded to it in the Council established in October 1999. His idea of a NSC was of an organization with a strong civilian representation and control. With the military's involvement in policy-making it would, however, have been difficult for an Army general to blame the political government alone for all the wrongs in the country. The Prime Minister rejected the idea not because it was originally raised by a military dictator, General Zia-ul-Haq (after all, Sharif was also a product of the Zia regime, personally groomed by the general), but because it would also have been tantamount to sharing power, which Sharif was unwilling to do.

One opinion was that Sharif's removal was a consequence of the humiliation of the Army's failure in Kargil. This approach did not consider that Sharif's post-Kargil diplomatic efforts actually saved the military, even the entire nation, from a disaster and further humiliation. The military operation had not taken into consideration a number of factors, such as the country's poor economic state. Furthermore, while planning such an operation the Army did not take into account the fact that such a venture cannot be launched without the concerted efforts of the military, the government and its diplomatic machinery. The Foreign Office, in particular, was left far behind, hence, it was not prepared to withstand the international pressure that followed the Kargil operation. The reality is that the removal of the government was more about saving the interests that Sharif had threatened. After the Kargil operation it was difficult to justify the logic of a large standing military. In fact, the government had asked the armed forces to reduce personnel substantially but any prospective downsizing was detrimental to the interests of the generals. Sharif's expulsion indeed reinforced the military's position in defense policy-making.

Understandably, some might interpret the latest coup as the death of democracy in Pakistan. This might be the case in the short-term, but in the long-term the democratic process is likely to emerge stronger than in the past. It may be the last time that an Army general is able to impose arbitrary rule or felt it incumbent upon him to do so. There are two explanations for this argument: first, if the Army leadership really set itself upon improving the quality of the democratic process in the country through devolution of power to the grass-roots level and enforcing the much needed accountability in the country, democracy would emerge much stronger than ever before. The moral, financial and political corruption of the political leadership has been an ugly reality of Pakistani politics and any effort at improvement would bear fruit. Second, if the military government fails to deliver the results it has promised, it would generate resentment against the most powerful institution in the country. The reaction, which may be violent, would naturally isolate the Army even further and would lead to a process of downsizing and curtailment of its authority. A word of caution for those who believe that the military leadership would be able to introduce accountability in the country and improve the democratic process is that the concept of accountability is alien to the Army. A major change in Pakistan, nevertheless, would be a long-term objective, and until that happens, the military is likely to remain in charge of defense decision-making.

The economic scene

Even with an absolute control over decision-making, military managers would find it hard to finance military modernization. In 1998–99 the armed forces were looking for about $10 billion to be spent over a span of seven years for the procurement of conventional hardware. This was in addition to the funds needed for the weaponization of nuclear deterrence. Given the poor state of the national economy it would be difficult to find these resources.

Economic conditions worsened to a degree that Islamabad was unable to meet its financial commitments to economic aid donors. In September/October 1999, the foreign exchange reserves were as low as $1.3 billion. Exports had dwindled to $5653 million in 1998/99 *vis-à-vis* the 1997–98 figures of $8434 million and foreign direct investment had reduced to $300.7 million in 1998–99 compared with $822.6 million in 1997–98. The trade balance, in any case, was in deficit. In the first quarter of FY 1999–2000 the trade deficit had increased to $369 million. Although imports showed a downward trend of –11.1 per cent (July 1998–March 1999), the real decline was in capital goods imports, which went down in 1997–98 to 31 per cent of total imports from 33 per cent in 1990–91. Consumer goods, on the other hand, showed an increase rising up to 20 per cent in 1997–98 from 16 per cent in 1990–91.[1] A decline in capital goods imports, mainly comprising industrial machinery, meant less industrial growth and related activities. This, in itself, was ominous for economic growth. In fact, industry on the whole experienced a downward trend with growth of 3.3 per cent in 1996–97, as opposed to 4.6 per cent in 1992–93. Large-scale manufacturing was the most affected with the net value declining from Rs. 69 039 million in 1995–96 to Rs. 68 051 million in 1996–97. The agriculture sector was no different, showing no positive trends either. Major crops production decreased by 2.2 per cent in 1996–97. The GDP also declined from 6.7 per cent in 1977–88 to 4.3 per cent in 1988–98.[2] The general index for share prices continued to show a downward trend since 1992–93. A similar poor performance could be seen in other areas as well, for example, capital formation and fiscal deficit. In the first instance, total investment decreased to –4.52 per cent of GDP.[3] This figure, however, did not factor in the inflation and therefore the real investment rate would be much lower than the official figure. In 1998–99 the government had a deficit of about nine per cent of the GDP which was not a good sign for a government that owed approximately $31 billion to foreign donors including the World Bank and the IMF.

This deterioration had certainly not come about in days. There were basic structural problems with the economy that no leadership had tried to solve, one of the major problems relating to the dependence upon external financial resources. The government increasingly started to depend upon workers' remittances and foreign economic assistance to balance its accounts. This had a paralysing impact on economic planning. After 1985, economic policies primarily aimed at short-term financial survival of the state. With the balance of payment declining negatively, Islamabad resorted to the practice of borrowing funds on commercial rates of interest, a causal factor in the ballooning of the debt burden. Under these circumstances, it was natural for the government to feel the pressure when foreign remittances almost dried up after the nuclear tests in July 1998.

The economic emergency imposed by Islamabad after the nuclear tests, and the decision not to honor commitments made with International Power Producers (IPPs) led to the loss of investors' confidence and a decline in private capital inflows. In the first half of FY 1998–99 official external reserves declined to $450 million (three weeks of imports) and external payment arrears accumulated to $1.4 billion on both current and capital accounts despite a substantial compression in imports. The economy had become virtually dependent upon the IMF loan for survival. The Fund's financial assistance, however, was linked with the government's ability to carry out structural adjustments. The donor's greatest objection was that efforts were not being made to enhance the tax net. Despite repeated promises, Islamabad had failed to impose the general sales tax (GST). This was mainly due to the negative policies of the government. There were massive strikes and disturbances after the announcement of the imposition of the GST. The business community, which Sharif seemingly represented, had refused to support him in his action. It was feared that GST would result in a price hike that would lead to a further decline in business. The business community was also not willing to share the burden of the results of financial mismanagement of the establishment. Financial mismanagement, in fact, was a major reason for the erosion of faith in the government's policies. Sharif's unhealthy investment policies even made the foreign aid donors extremely nervous and highly skeptical of the economic problems faced by the country.

The new Army regime hopes to find a breakthrough in improving the financial conditions by recovering national wealth looted by corrupt politicians and businessmen in the shape of loans, hoping thereby to revive investors' confidence in the country. The strategy denotes a well-meaning, albeit a short-term approach to improve the economic health of the country. By the end of 1999 there were no signs of a stronger plan to put the economy, including infrastructure development, investment, industry, exports and other sectors, back on track.

The poor economic performance will continue affecting the defense sector as much as other sectors. In one respect Sharif's policies were contradictory. On the one hand he supported the military's desire to counter India's hegemonic designs, but on the other he was unable to put the economy on a track that would have enabled the armed forces to carry out military modernization plans. Most of the major weapons acquisitions plans were halted by the end of 1998 because of a dearth of funds. Nuclear deterrence itself demands commitment of resources. Some people proposed that these resources should be made available by reorganizing and restructuring the armed forces, hence, reallocating funds[4] but there were some inherent problems in this argument. First, the funds released through reorganization will not match exorbitant spending on nuclear weapons. Second, the lack of expertise in the defense sector would not allow restructuring to take place. Would, then, the establishment divert resources from the social sector? That could be considered as an option but with serious implications, none the less. The political cost of further diversion of resources will be more than what the leadership can afford. The military may have to depend upon the only other significant source, its corporate ventures, hence, in the past, profits were diverted to meet military requirements. The mushroom growth of military controlled corporate ventures such as real estate, aviation, security services, banking, construction, commercial airline and other heavy industrial units, with no accountability, have and will provide opportunity to the armed forces to foot some of its modernization bills. The resources generated thus would, however, be insufficient to cater to major capital investments. What the potential is of these huge business conglomerates is a question that requires further research. The fact is, unless the economy gets back on its feet it would be difficult for any government to support the kind of military buildup envisioned by military managers. The other option is somehow to force New Delhi to slow down its nuclear technological advancements. This would ensure the military capabilities and balance remain manageable until the Pakistani economy recovers to a degree where a major weapons modernization can be undertaken again.

The strategic scene

The 1990s saw the intensification of hostilities with India. Antagonistic relations with its neighbor were a focal point of Islamabad's security framework. Pakistan had lost two wars against its adversary and almost a third defeat in 1965. The memories of wars and humiliating defeats were instrumental in sustaining military buildup. A corollary of this was the understanding, which has implanted in the minds of military managers over the years, that well prepared military tactical and operational plans, along with acquisition of superior technology, were a prerequisite for keeping New Delhi's hegemonic designs at bay. Not only this, military superiority could allow Pakistan to force India to let go the disputed territory of Kashmir. Islamabad's military ambitions went beyond self-defense. The majority of the policy-making elite hoped to humble India's desire to project itself as a regional power.

Nuclear proliferation conveniently fitted into this style of thinking. With this non-conventional technology, possessed by a few countries in the world, Islamabad could force its traditional adversary to reopen the Kashmir case and possibly solve it to Islamabad's taste. This was indeed the idea behind the Pakistan Army's initiative at Kargil. Planned as a fine but limited tactical manoeuvre, this operation could not provide the required dividends precisely because of the effect of the nuclear option to South Asian strategic culture. It was, somehow, hoped that India, deterred by Islamabad's nuclear capability and pressurized by its tactical victory at Kargil, would concede to solving the Kashmir issue. Indubitably, this thinking was not logical since the plans did not take into account a number of other issues such as the country's economic capacity to withstand a war. The pressure on India would have to be exerted diplomatically and the process of capitalizing upon a military manoeuvre to gain long-term political dividends was necessarily protracted. Pakistan was not prepared for this. The government's diplomatic machinery was too de-motivated and directionless to take on India diplomatically. There was, also, a communication gap among the Army, the Foreign Office, and the political leadership that made a tactical victory ineffective before very long. Consequently, India managed to divert the pressure on Pakistan instead. In doing so, New Delhi capitalized upon the fear of the international community that if it ever came to a war between the two, Pakistan was more likely to switch to a nuclear option so, hence, Islamabad must be stopped from pursuing its military agenda. The Pakistan Army's strategy had obviously backfired. India proved more resilient in sustaining diplomatic pressure. The diplomatic battle finally made Islamabad retreat. The tactical victory followed by a strategic defeat at Kargil denoted the limitations of nuclear deterrence in effecting a military oriented solution of a contentious issue. In one respect, the event also terminated any further debate on the issue between India and Pakistan. New Delhi was always adverse to the idea of considering Kashmir as a disputed territory, its contention being that Kashmir was an internal matter and did not require any foreign mediation or external interference. Contrastingly, Islamabad desires to solve the issue through multilateral negotiations. The closest that New Delhi came to conceding to Islamabad's ideas was to agree to bilateral discussions. This chapter, however, was closed with the Pakistan Army's venture in Kargil. It does not appear likely that Islamabad would be able to reopen the issue at least in the near future. If a solution to the Kashmir issue has to come at all, it would be due to sustained pressure from the Kashmiris and an Indian government's desire to solve a painful problem. Would this new situation mean the Pakistan military changing its strategy and dropping Kashmir from its list of strategic priorities? This did not appear likely, especially after the military's takeover.

Washington, on whom Islamabad had depended, was averse to the idea of Pakistan disturbing the *status quo* in the region. This American reaction was perceived by some as Washington's bias against an Islamic country becoming a nuclear power as opposed to a Hindu nation attaining this capability. Samuel Huntington's theory of the clash of civilizations was certainly lapped up by a number of people. This theory was relevant to the ideas of those uncomfortable with Pakistan's unfavorable placement in the global system. The security culture in South Asia, particularly in Pakistan, had increasingly become focused on the paranoia of Pakistan being relegated to second place in the region. Attaining a position through challenging existing hierarchical systems was viewed as a way of climbing up the ladder. Increasingly in the military, people subscribed to the idea of Pakistan emerging as a leader of an Islamic bloc. The ISI's support of religious extremists, especially in Afghanistan, was detrimental to American interests. For the *Talibaan* and other religious extremists, Afghanistan has become a safe haven to pursue terrorist activities aimed primarily against the US. This was an issue that seemingly bothered American policy-makers more than did Kargil. In September 1999 the US State Department started negotiations with Pakistan to desist from supporting religious extremists. The civilian government seemed to have agreed to cooperate, as could be observed from the statements made by a number of senior government functionaries condemning Afghanistan as a breeding ground for terrorism. Between the humiliating retreat from Kargil and pressure on Pakistan from a number of sources such as America, Iran, Russia and the CARs to stop supporting terrorism, Islamabad did not have many options. Unlike in the past, military managers would have to rethink their Afghan policy and withdraw support from the *Talibaan*. For an Army government, invoking additional pressure from the international community on account of its questionable Afghan policy, would not be desirable.

To develop a worthwhile dialogue with the world, the Pakistani generals necessarily have to review and revise their military posture. For the time being, this idea mystifies the policy-makers who believe that military prowess is essential for the country's survival, despite the grave internal security problems. Towards the end of the century the greatest threat to Pakistan is purely internal in nature, the rampant corruption and mismanagement of national resources posing a much more formidable threat. Furthermore, the ethnic and sectarian differences leading to violence is a frightening reality. The society has grown to be militaristic, a development in which the state has a fair share. The erection of models of indigenous ballistic missiles such as 'Ghauri' and 'Shaheen' in almost every city nurtures a peculiar psyche based on conflict rather than cooperation or peaceful coexistence domestically or internationally. Thus, any external debate on reducing defense spending, or non-interference in Afghanistan was construed as intervention in the country's internal affairs, leading to further isolation within the international framework. Indubitably, the process of isolation began with Islamabad getting less support from both the Islamic world and traditional allies like China. Nawaz Sharif's efforts to sell his country's image as the only nuclear capable Islamic country failed but it was somehow hoped that other Islamic countries would come to Pakistan's financial assistance in recognition of its non-conventional defense capability. This was an erroneous assumption. It was like the Indian psyche with a difference. While India has been equally possessed with the desire to be accepted militarily as a regional power, Pakistan's efforts were directed towards an extra-regional arrangement. It was hoped that Islamabad would be accepted as the leader of an Islamic bloc but Pakistan had little chance in attaining any political or strategic significance within the South Asian region. Strengthening the military posture, therefore, was the fundamental objective. To attain this, a two-pronged approach was adopted: (a) to compete militarily with India and (b) align itself with powers such as the US and China. While relations with the United States underwent a decline from the end of the 1980s, bilateral ties with China were sustained, albeit subtly changed. China no longer considered a strategic alliance with Pakistan as one of its priorities. Although Beijing was one of the major sources of military technology procurement, and it would continue to provide the equipment to Islamabad in future, China was not willing to take the risk of building relations with its South Asian ally at the cost of excessively antagonizing India.

At the end of the century, Pakistan appears to have exhausted its options. Its military posture and related defense buildup has deprived it of a number of opportunities for developing the country's socioeconomic conditions. A continuation on these lines is no longer an easy task. Irrespective of what form of regime rules the country, the leadership would have to weigh its options for continued military buildup versus disarmament. At the very least, policy-makers must seriously rationalize military planning and reduce the defense burden through improving management of the defence sector. Rational plans include abstaining from engaging in a suicidal military competition with India. This would have to be done despite the BJP's return to power in India, a party reputed to have extremist views. New Delhi is likely to singularly pursue the objective of turning India into a military power on the Asian continent, a policy that would stop it from negotiating peace and resolving outstanding issues with traditional foes such as China and Pakistan. Does Pakistan wish to respond to India in the same way? More important, can Pakistan afford to compete with its traditional adversary militarily? The prospect of continued military competition with India is extremely hazardous for Pakistan and would lead to further retardation of socioeconomic development, the country's greatest need. Moreover, it must not be forgotten that, at this juncture, national resources can no longer sustain the burden of fulfilling military modernization goals.

Notes

Introduction

1 Ian Anthony, *The Arms Trade and Medium Powers: Case Studies of India and Pakistan, 1947–1990* (London: Harvester Wheatsheaf, 1992), pp. 21–81.
2 Ibid. pp. 137–45.
3 Ibid. pp. 7–8, 100.
4 Ibid. pp. 5, 8, 16–17.
5 Pervaiz Iqbal Cheema, *Pakistan's Defense Policy, 1947–58* (London: Macmillan, 1990), pp. 105–55.
6 Saadet Deger and Somnath Sen, 'Military Security and the Economy: Defense Expenditure in India and Pakistan', in Keith Hartley and Todd Sandler, *The Economics of Defense Spending* (London: Routledge, 1990), p. 200.
7 Robert G. Wirsing, *Pakistan's Security under Zia, 1977–1988* (London: Macmillan, 1991), pp. 81–142.
8 Akhtar Ali, *Pakistan's Nuclear Dilemma* (New Delhi: ABC Publishing House, 1984), pp. 1–20, 61–87. See also, Ian Anthony, *op. cit.,* pp. 153–69. Shlomo Aronson with Oded Brosh, *The Politics and Strategy of Nuclear Weapons in the Middle East* (New York: State University of New York, 1992), p. 261. Muhammad Zafar Iqbal Cheema, *Indian Nuclear Strategy, 1947–91* (London: PhD thesis submitted to the University of London, August 1991), pp. 155–8. Martin Van Creveld, *Nuclear Proliferation and the Future of Conflict* (New York: Free Press, 1993), pp. 85–6. Zeba Moshaver, *Nuclear Weapons Proliferation in the Indian Subcontinent* (London: Macmillan, 1991), pp. 60–76, 153–4.
9 Ashok Kapur, *Pakistan's Nuclear Development* (London: Croom Helm, 1987), pp. 100, 108–09, 194–5.
10 Hasan-Askari Rizvi, 'Factors Shaping Pakistan's Arms Production Programme', *Strategic Digest* (New Delhi: May, 1989), pp. 538–42. See also Michael Brzoska and Thomas Ohlson, *Arms Production in the Third World* (Stockholm: SIPRI, 1986), pp. 264–8. Nazir Hussain, *Defense Production in the Muslim World* (Karachi: Royal Book Company, 1989), pp. 33–45.
11 Ron Mathews, Country Survey IV: Pakistan *Defense and Peace Economics* (York: V, 1994).

1 Development of Threat Perception

1 Interview with Lt General (Retd.) Hameed Gul and Lt General (Retd.) Kamal Matinudin (Rawalpindi: 15/05/94 and 23/02/94).
2 Rodney W. Jones, 'The Military and Security in Pakistan', in Craig Baxter (ed.) *Zia's Pakistan: Politics and Stability in a Front Line State* (Lahore: Vanguard, 1985), p. 34.

3   Interview with Professor Stephen Cohen (Nathiagali: 16/05/94).
4   Raja Anwar, *The Terrorist Prince* (Lahore: Vanguard, 1998), p. 41–5.
5   Oliver Roy, *The Lessons of the Soviet/Afghan War* (London: IISS, Adelphi Paper 259. Summer 1991), p. 3.
6   Anatoly Dobrynin, *In Confidence* (New York: Random House, 1995), p. 434–43.
7   Interview with the Director of IDSA, India, Air Commodore (Retd.) Jasjit Singh (London: 15/02/95).
8   Louis Dupree, 'Pakistan and the "Afghan Problem"' in Craig Baxter and Syed Raza Wasti, *Pakistan Authoritarianism in the 1980s* (Lahore: Vanguard, 1991), p. 59.
9   Gulf Crisis 1990, Rawalpindi: Army GHQ Publication. For reference see the *Times* (London: 14/01/92).
10 S.M. Burke, and Lawrence Ziring, *Pakistan's Foreign Policy: An Historical Analysis* (Karachi: Oxford, 1990), p. 460.
11 Selig S. Harrison, and Geoffrey Kemp, *India & America. After the Cold War* (Washington, D.C.: Report of the Carnegie Endowment Study Group on US–Indian Relations in a Changing World Environment, 1993), p. 19.
12 Zafar Iqbal Cheema, *India's Nuclear Strategy, 1947–91* (London: PhD thesis presented at the Department of War Studies, King's College, University of London, 1991), p. 235.
13 Raju G.C. Thomas, 'The Growth of Indian Military Power: From Sufficient Defense to Nuclear Deterrence', in Ross Babbage and Sandy Gordon, *India's Strategic Future. Regional State or Global Power?* (London: Macmillan, 1992), pp. 41–3.
14 Mehnaz Isphani, *Pakistan: Dimensions of Insecurity* (London: IISS, Adelphi Paper 246. Winter 1989–90), p. 30.
15 Interview with Brig (Retd.) Noor Hussain (Rawalpindi: 09/11/93).
16 Robert G. Wirsing, *India, Pakistan and the Kashmir Dispute* (London: Macmillan, 1994), pp. 208–11.
17 Interview with Director IDSA, India, Air Commodore (Retd.) Jasjit Singh (London: 15/02/95).
18 Interview with Lt General (Retd.) Rahim–u-Din Khan (Rawalpindi: 07/02/94).
19 *Ibid.* This idea was also subscribed to by the former Chief of General Staff, Lt General (Retd.) Farrukh Khan (Rawalpindi: 29/03/94).
20 Interview with the Director of IDSA, India, Air Commodore (Retd.) Jasjit Singh (London: 15/02/95).
21 Interview with the then DG (Military Operations)-Pakistan, Maj. General Prevaiz Musharaf (Rawalpindi: 29/03/94).
22 Interview with Lt General (Retd.) Hameed Gull (Rawalpindi: 15/05/94).
23 Zafar Iqbal Cheema, *op. cit,*, p. 199.
24 *Foreign Affairs*, **XXXIX** (7), (July 1990), p. 91.
25 *Strategic Analysis,* (March 1990), pp. 1274–7.
26 *Jane's Defence Weekly* (London: 2 December 1989), p. 1230.
27 Chandra B. Khanduri, 'The Zarbe Momin and Changed Pak Perspective: An assessment', *Strategic Analysis,* **XII** (7), (New Delhi, March 1990), p. 1277.
28 K. Subrahmanyam, 'Pak Offensive–Defensive Talk is Premature'. *Times of India,* (31/12/89).

29 Interview with Lt General (Retd.) Alam Jan Mehsood (Rawalpindi: 22/02/94).
30 John Kaniyalil, 'Zarbe Momin: Pakistan's Show of Offensive Defense', *Foreign Affairs Reports* XXXIX (7), (July 1990), p. 86.
31 See discussion in the Henry L. Stimson Center Occasional Paper No. 17, (July 1994), p. 21.
32 *Bulletin of Atomic Scientists* (07/08/92), p. 39. See also *Far Eastern Economic Review*, (30/04/92), p. 25.
33 Alastair Lamb, *Kashmir. A Disputed Legacy, 1846–1990* (Hertfordshire: Roxford, 1991), pp. 322–41.
34 *Times of India* (11/01/90).
35 Robert G. Wirsing, *op. cit.*, p. 155.
36 Interview with Maj. General Muhammad Safdar (Lahore: 03/04/94). (He was GOC at that time.) See also, Robert G. Wirsing, *op. cit.*, p. 122.
37 Interview with Dr Mehboob–ul–Haq (Islamabad: May 1997).
38 Hasan-Askari Rizvi, *Pakistan and the Geostrategic Environment* (New York: St Martin's, 1993), p. 39.
39 Ravi Rikhye, *The Militarisation of Mother India* (New Delhi: Prism, 1990), p. 34.
40 Robert G. Wirsing, *Pakistan's Security Under Zia, 1977–88* (London: Macmillan, 1991), p. 89.
41 Robert G. Wirsing, 'Pakistan's Security Predicament', *Defense Journal* (Karachi: August, 1985), p. 17.
42 Raju G.C. Thomas, *op. cit.*, p. 52.
43 *The Military Balance, 1981–82* (London: IISS, 1981), pp. 80, 86.

2 The Official Decision-Making System

1 General (Retd.) Mirza Aslam Baig, *Restructuring the Higher Defense Organisation* (Rawalpindi: Friends, 1993) p. 2.
2 General (Retd.) Muhammad Musa, *My Version: India-Pakistan War, 1965* (Lahore: Wajid Alis, 1983), pp. 2–3. See also, Sir James Morrice, *Pakistan Chronicle* (Karachi: Oxford, 1993), p. 132.
3 General (Retd.) Mirza Aslam Baig *op. cit.*, p. 3.
4 Pakistan's White Paper on Higher Defense Organisation, *Strategic Digest* (New Delhi: June, 1976).
5 Brig. (Retd.) Karrar Ali Agha, 'Concept of Supreme Command: A Dangerous Duality-II', *Nation* (Lahore: 04/12/89).
6 See article 243 (1A) of the 1973 Constitution of the Islamic Republic of Pakistan. See also, Ian Anthony, *The Arms Trade and Medium Powers. Case Studies of India and Pakistan, 1947–1990* (London: Harvester Wheatsheaf, 1992), pp. 17.
7 Franklyn A. Johnson, *Defense by Ministry* (London: Duckworth, 1980), pp. 54, 107.
8 David M. Lampton, 'A Plum for a Peach: Bargaining, Interest and Bureaucratic Politics in China', In Kenneth G. Lieberthal, and David M. Lampton (eds), *Bureaucracy, Politics and Decision Making in Post-Mao China* (Berkeley: University of California, 1992), p. 9.

9   David S. Sorenson, 'Fewer Dollars and More Reform? Defense Acquisition in the United States after the Cold War', *Security Studies*, II (2), (Winter 1992).
10 Interview with General Shamim Alam Khan (Rawalpindi: 21/03/94).
11 The Combat Development directorate responsible for evaluating weapons technology was re-designated in 1998–99 as the Directorate of Strategic Development.
12 Interviews with DCAS (Operations) Air Marshal Abbas Khattak (Rawalpindi: 18/05/94), and DCNS (Operations) Rear Admiral Obaidullah Khan (Islamabad: 16/04/94). See also, *The Story of the Pakistan Air Force* (Islamabad: Shaheen Foundation, 1988), pp. 13–14.
13 E.E. Chaffee, 'Three Models of Strategy' *Academy of Management Review*, (10/1985), p. 1.
14 David Braybrooke and Charles E. Lindblom, *A Strategy of Decision: Policy Evaluation as a Social Process* (London: Macmillan, 1970), pp. 29–61.
15 Agha Shahi, *Pakistan's Security and Foreign Policy* (Lahore: Progressive, 1988), p. 218.
16 Interview with Lt General (Retd.) K.M. Arif (Rawalpindi: 14/11/93).
17 See the section on defense in the *Annual Compilation of Appropriation Accounts, 1980–92* (Islamabad: Ministry of Finance).

3 Pakistan's Power Politics and Defense Decision-Making

1   Mehnaz Isphani, *Pakistan: Dimensions of Insecurity*, Adelphi Papers 246 (London: IISS, Winter 1989–90), p. 32.
2   *The News* (Lahore: 15/02/88).
3   Gavin Kennedy, *The Military in the Third World* (New York: Charles Scribner's, 1974), p. 89.
4   Kamal Azfar, 'Constitutional Dilemmas in Pakistan', in Shahid Javed Burki and Craig Baxter (eds) *Pakistan Under the Military: Eleven Years of Zia ul-Haq* (Boulder: Westview, 1991), p. 69.
5   These are schools that exist only on papers with no physical existence and are a clear example of corruption in the establishment.
6   Interview with Lt General (Retd.) Hameed Gull (Rawalpindi: 15/05/94).
7   Interview with Moen Aftab (Rawalpindi: 08/02/94).
8   Ian Anthony, *The Arms Trade and Medium Powers. Case Studies of India and Pakistan, 1947–90* (London: Harvester Wheatsheaf, 1992), p. 19.
9   Ayesha Jalal, *The State of Martial Rule* (Lahore: Vanguard, 1991) and Gavin Kennedy, *op. cit.*
10 Interview with Lt General Muhammad Sharif (Rawalpindi: 24/02/94).
11 Interview with Lt General (Retd.) Muhammad Iqbal (Rawalpindi: 28/02/94).
12 Interview with Admiral (Retd.) Iftikhar Sirohey (Islamabad: 27/02/94).
13 Interview with General Shamim Alam Khan (Rawalpindi: 21/03/94).
14 Mushahid Hussain, *Pakistan's Politics: The Zia Years* (Lahore: Progressive, 1990), pp. 155–6.
15 Source: Deputy Director Audit (Defense Procurement).
16 Interview with the Commander Pakistan Fleet. Rear Admiral (Retd.) Shamoun Alam (Karachi: 03/05/94).
17 Interview with Moen Aftab (Rawalpindi: 08/02/94).

18 Christopher Smith, 'The Topography of Conflict: Internal and External Security Issues in South Asia', *Brasseys Defence Yearbook 1993* (London: Brassey's, 1993), p. 302.
19 Sreedhar (ed.) *Pakistan's Bomb – A Documentary Study* (New Delhi: ABC, 1986), p. 107.
20 Interview with Dr A.Q. Khan (Islamabad: 28/02/94).
21 Zahid Malik, *Dr. A.Q. Khan and the Islamic Bomb* (Islamabad: Hurmat, 1992), p. 128.
22 Kausar Niazi, *Last Days of Premier Bhutto* (Lahore: Jung, 1991), pp. 88–9.
23 Interview with Additional Secretary Defense, Lt General Jamshed Malik (Rawalpindi: 10/02/94).
24 See Aroosa Alam 'Law to Check Defense Kickbacks Likely', *The Muslim* (Islamabad: 30/08/91).
25 Interview with Riaz Paracha (Islamabad: 01/12/93).
26 Interview with the former Foreign Secretary. Shehryar Khan (Islamabad: 15/05/94).
27 Samina Yasmeen, 'Pakistan's Cautious Foreign Policy', *Survival* (London: IISS, Summer 1994), pp. 121–6.
28 *Audit Report Defense Services Vol-IV, 1988–93* (Islamabad: Auditor-General of Pakistan).
29 See Ikram Sehgal, 'The Last of the Mohicans!' *Nation* (Lahore: 11/12/91).
30 Munir Ahmed, *Political Role of Intelligence Agencies in Pakistan* (Urdu) (Lahore: Jahangir, 1993), pp. 74–101, Appendices 1–9.
31 Ayesha Jalal, *Democracy and Authoritarianism in South Asia* (Lahore: Sange Meel, 1995), pp. 144–7.
32 Interview with Lt General (Retd.) Hameed Gull (Rawalpindi: 15/05/94).
33 *Herald* (Karachi: April 1994), pp. 25–32.
34 Interview with General (Retd.) Mirza Aslam Baig (Rawalpindi: 15/05/94).
35 Robert G. Wirsing, *India, Pakistan and the Kashmir Dispute* (London: Macmillan, 1994), pp. 120–4.

4 The Cost of Military Buildup

1 Saadet Deger and Somnath Sen, 'Military Security and the Economy: Defense Expenditure in India and Pakistan', in Keith Hartley, and Todd Sandler (eds) *The Economics of Defense Spending. An International Survey* (London: Routledge, 1990), pp. 198–219.
2 Omar Noman, *Pakistan: Political and Economic History Since 1947* (New York: Kegan Paul, 1988), p. 84.
3 This was stated by Dr Mahbub ul Haq during his keynote address at the conference on 'Converting Defense Resources to Human Development' organized by the Bonn International Center for Conversion 9–11 November, 1997.
4 The Army has approximately 650 000 workers compared with 50 000 in the Air Force and 30 000 in the Navy.
5 This not only relates to kickbacks in major arms deals but other expenditure as well. For example, in FY 1990–91 the PAF was given a budget of Rs. 5.4 million as a travel rebate for PAF officers for air and rail (military

officers are provided with 50 per cent discount on air and rail travel). The remainder is paid by the services, hence resources are allocated for the purpose. It is also a fact that officers tended to misuse the facility but the government did not take any action to reduce the spending. At the end of the year, an amount of Rs. 3.5 million was saved in this head of account. Under government financial rules, the budget for the following year is made on the basis of the previous year's budget. A non-usage of funds has an impact on the following year's allocation. Resources are deducted accordingly. None the less, this rule was not adhered to in allocating resources for PAF's expenditure. Under this particular head allocation was consistently increased from Rs. 8 million in FY 1991–92 to Rs. 17.9 million in FY 1994–95.

6 The Askari Bank was rated as the top private bank with the biggest funds reserve in the country.

7 For instance, the Bahria Foundation was given the contract for the sale of spares that were acquired from public funds for the eight frigates leased from the US in 1988. It is scandalous that the procuring agencies continued to order spares for these frigates even when it had become clear that the ships were to be called back and spares were received after the frigates returned to Washington. Later on, the foundation was given the contract to sell the spares with an understanding that it would keep 50 per cent of the sale proceeds. Moreover, these were sold at a lower price than what the government would have managed resulting in a loss of millions of rupees to the treasury.

8 Interview with the former Finance Minister, Sirtaj Aziz (Islamabad: 19/10/97).

9 Charles A. Kupchan, 'Defense Spending and Economic Performance', in *Survival* (London: IISS, September–October 1989), pp. 450–1.

10 Mahbub ul Haq, *Human Development in South Asia, 1997* (Islamabad: Oxford, 1997).

11 *Ibid.* p. 18.

12 Interview with several people from the education sector (Islamabad: 15/05/98, 20/05/98 and 13/02/99).

13 Mushahid Hussain, 'Pressure put on Pakistani Spending' *Jane's Defence Weekly* (London: X (2) 16/07/88), p. 70.

14 *Ibid.*

15 Interviews with the Division Chief (Policy Development and Review Department), Roger H. Nord, and economist, Khaled I. Sakr (Washington: 07/12/97).

16 Interview with economist of the World Bank in Pakistan, Mr Ghulam Qadir (Islamabad: 20/04/94).

5 Pakistan's Arms Suppliers

1 Hasan-Askari Rizvi, *Pakistan and the Geostrategic Environment* (New York: St Martin's, 1993), p. 88.

2 Interview with Lt General (Retd.) K.M. Arif (Rawalpindi: 14/11/93). He was part of the Pakistani negotiation team.

3 According to Professor Stephen Cohen, working in the State Department at the time of the Soviet invasion of Afghanistan, opinion in the American Congress was divided. Twenty per cent of members thought that the USSR's objective was merely the invasion of Afghanistan; 60 per cent thought it was to move closer to the Gulf, and only 10–20 per cent were of the view that there were plans to invade Pakistan. It was afterwards that thinking in the US Congress started to change, giving place to the idea that Pakistan's security was directly threatened. Interview (Nathiagali: 16/05/94).

4 Rodney W. Jones, 'The Military and Security in Pakistan', in Craig Baxter (ed.), *Zia's Pakistan: Politics and Stability in a Front Line State* (Lahore: Vanguard, 1985), p. 34.

5 Pandak Nayak, 'Pakistan's Gulf Connections in Perspective', in Akhtar Majeed, *Indian Ocean: Conflict & Regional Cooperation* (Lahore: Ayaz, 1987), p. 118.

6 Congressional Record-Senate (Washington: 03/03/83), pp. S2062–5.

7 *Ibid.*, pp. S2052–65.

8 S.M. Burke and Lawrence Ziring, *Pakistan's Foreign Policy: An Historical Analysis* (Karachi: Oxford, 1990), p. 447.

9 Oliver Roy, *The Lessons of the Soviet-Afghan War*, Adelphi Paper 259, (London: Summer 1991), p. 36.

10 Nimrod Novik, 'Weapons to Riyadh: US Policy and Regional Security' (Tel Aviv: Center for Strategic Studies, April 1981), pp. 3–12.

11 Fred Halliday, *The Making of the Second Cold War* (London: Verso, 1986), pp. 46–70.

12 Interview with Lt General (Retd.) Ijaz Azeem (Islamabad: 09/12/93).

13 Mitchell Reiss, 'The Illusion of Influence: the United States and Pakistan's Nuclear Programme', *International Security* (London: RUSI Journal, Summer 1991), p. 47.

14 *Jung* (Lahore: 28/06/81).

15 Interviews with Lt General (Retd.) Muhammad Safdar (Lahore: 29/04/94) and the former foreign secretary, Niaz. A. Naik (Islamabad: 23/02/94).

16 Interview with Pervaiz Iqbal Cheema (Islamabad: 14/02/94).

17 Keith R. Krause and Andrew Latham. 'Constructing Non-Proliferation and Arms Control: The Norms of Western Practice', in Keith R. Krause, *Culture and Security* (London: Frank Cass, 1999), p. 41.

18 Leonards Spector, 'Neo-Non-proliferation', *Survival* (London: IISS, Spring 1995), p. 77.

19 *Herald* (Karachi: April 1994), p. 28.

20 Selig Harrison, and Geoffrey Kemp, *India and America: After the Cold War*, Report of the Carnegie Endowment Study Group on US-Indian Relations in a Changing World Environment (Washington: 1993), pp. 25–41.

21 Devin T. Hagerty, 'India's Regional Security Doctrine', *Asian Survey* **XXXI** (4) April, 1991, p. 363.

22 Raju G.C. Thomas, *South Asian Security in the 1990s*, Adelphi Papers 278 (London: IISS, July 1993), p. 11.

23 John Bray, 'New Directions in Pakistan's Foreign Policy', *World Today* (London: April 1992), p. 65.

24 *Jung* (Lahore: 09/03/95).

25 'The Gulf Crisis 1990' (Rawalpindi: ISPR, 1990). See reference to this paper in *The Times* (London, 14/01/92).
26 Interview with VCAS, Air Marshal Ali-u-Din (Rawalpindi: 17/02/98).
27 Samina Yasmeen, 'Chinese Economic and Military Aid to Pakistan, 1969–79', Working Paper No: 6, Department of International Relations (Canberra: Australian National University, January 1987), pp. 1–10.
28 Gary Klintworth, 'Chinese Perspectives on India as a Great Power', in Ross Babbage and Sandy Gordon, (ed.) *India's Strategic Future. Regional State or Global Power?* (London: Macmillan, 1992) p. 96.
29 Interview with former Foreign Secretary Shehryar Khan (Islamabad: 14/05/94).
30 Raju G.C. Thomas, *op. cit.*, p. 13.
31 Hasan-Askari, Rizvi *Pakistan and the Geostrategic Environment* (New York: St Martin's, 1993), pp. 148–62.
32 SIPRI database. Imports of Major Conventional Weapons by Pakistan, 1960–92.
33 David Shambaugh, 'Growing Strong: China's Challenge to Asian Security', *Survival* (London, Summer 1994), p. 53.

6 Military Industrial Complex

1 Message by General Abdul Waheed for the seminar on 'Self-Reliance in Defence' (Rawalpindi: MoD, 16–17/01/93), p. 11.
2 Hasan-Askari Rizvi, 'Factors Shaping Pakistan's Arms Production Programme', *Strategic Digest* (New Delhi: May 1989), pp. 540–1.
3 *The News* (Lahore: 04/09/91).
4 James Everett Katz, *Arms Production in Developing Countries* (Massachusetts: Lexington, 1984), p. 7.
5 Interview with the CGS, Lt General Farrukh Khan (Rawalpindi: 29/03/94).
6 *Far Eastern Economic Review* (15/11/90), p. 72.
7 Steven Nahmias, *Production and Operations Analysis* (Massachusetts: Irwin, 1993), p. 29. See also Tamir Agmon and Mary Ann Von Glinow, *Technology Transfer in International Business* (Oxford: Oxford, 1991), p. 89.
8 Edward J. Malecki, *Technology and Economic Development* (London: Longman, 1991), p. 123.
9 *World Development Report*, 1994. (Oxford: Oxford, 1994), p. 216.
10 Ron Mathews, 'Country Survey IV: Pakistan', *Defence and Peace Economics V* (York: 1994), p. 323.
11 Interview with Dr A.Q. Khan (Islamabad: 28/02/94).
12 Brig. Khalid M. Masud, 'Development of Strategic Materials and its Requirement for Defence', Paper presented at the seminar on 'Self-Reliance in Defence', organized by the Defence Production Division, MoD (Rawalpindi: 16–17/01/93), p. 72–3.
13 Interview with the Director of MIRDC, Dr IkramulHaq (Lahore: 03/04/94).
14 Ron Mathews, *op. cit.*, p. 321.
15 Interview with the Chairman, POFs Maj. General Mahmud Ali Durrani (Wah: 10/12/93).
16 *The Frontier Post* (Islamabad: 18/01/93).

17 Interviews with private sector industrialists (January–June 1994 and February 1997).
18 The organization of Director General (Defense Procurement) carries out purchases of raw materials centrally. This is in addition to what the facilities could procure themselves. The common practice was that materials are purchased at exorbitant rates and often there is duplication, which increases the cost of production tremendously.
19 Nathan Rosenberg, *Exploring the Black Box* (Cambridge: Cambridge, 1994), p. 13.
20 Interview with Director Munitions Production, Brig. Muhammad Baqir (Rawalpindi: 08/02/94).
21 Interviews with Shamshad Ali (Lahore: 04/04/94) and Rana Abdul Qadir (Lahore: 02/04/94). See also *The Muslim* (Islamabad: 24/05/91).
22 Interview with the Chairman POFs, General Mahmud Ali Durrani (Wah: 12/02/94).
23 *International Defense Review* (June, 1985), p. 939.
24 Interview with various private small arms sellers.
25 *The Muslim* (Islamabad: 14/02/93).
26 *The Frontier Post* (Lahore: 18/07/91).
27 C.H. Kirkpatrick, and F.I. Nixon, *The Industrialisation of Less Developed Countries* (Manchester: Manchester University, 1983), p. 14.
28 Interview with Maj. General (Retd.) Muhammad Riaz (Lahore: 02/02/94).
29 *The News* (Lahore: 13/09/89).
30 Interview with Maj. General (Retd.) Atta Muhammad Utra (Rawalpindi: 21/02/94).
31 Interview with Maj. General (Retd.) Farhat Ali Burki (Rawalpindi: 22/02/94).
32 Interview with Maj. General Ahmed Ali (Rawalpindi: 08/02/94).
33 Interview with Lt General Tanveer Naqvi (Karachi: 26/01/95).
34 These employees comprise one PhD, two MPhil, about ten MSc, 20–24 BSc, 8 BE, 20 Diploma holders and a large number of unskilled staff.
35 Interview with Director Munitions Production, Brig. Muhammad Baqir (Rawalpindi: 08/02/94).
36 *Jane's Military Vehicles and Logistics* (London: Jane's, 1994–95), p. 439.
37 Interview with John Begg (Karachi: 02/05/94).
38 Interview with Lt General (Retd.) Talat Masood (Islamabad: 02/11/93).
39 Interview with Air Vice Marshal Dilawar (Kamra: 22/12/93).
40 Interview with Air Vice Marshal Saeed Anwar (Rawalpindi: 29/07/97).
41 PAC Brochure. See also *All the World's Aircraft. 1992–93*, (London: Jane's, 1993), p. 172.
42 Ali Abbas Rizvi, 'PAC KAMRA: A Success in Self-Reliance', *Globe* (Karachi: March, 1994), p. 40
43 Interview with Air Vice Marshal Dilawar (Kamra: 22/12/93).
44 *Defense and Foreign Affairs Weekly*, (13–19/03/89) p. 3.
45 Interview with the Managing Director KARF, Air Commodore Naqvi (Kamra: 22/12/93).
46 Interview with Air Vice Marshal (Retd.) Yusuf Khan (Peshawar: 28/03/94).
47 Interview with the General Manager, Naval Dockyard (Karachi: 07/05/94).
48 Interview with Chief Scientist, DESTO Saleem Mehmood (Rawalpindi: 03/03/94).

49 Interview with Brig. Ghulam Sarwar (Rawalpindi: 02/03/94). See also 'MVRDE and its Role and Contributions in Defence Production', *Defence Journal*, (July/August, 1992), pp. 20–1.
50 Interview with Director ARDE, Brig. Muhammad Sarwar (Rawalpindi: 26/02/94).

7 Military Buildup Decisions, 1979–90

1 Interview with Air Chief Marshal (Retd.) Anwar Shamim (Islamabad: 28/02/94).
2 Interview with Lt General (Retd.) K.M. Ariff (Rawalpindi: 14/11/93).
3 Interview with the representative of Thomson-CSF in Pakistan (Islamabad: 01/02/94).
4 Duncan S. Lennox (ed). *Jane's Air-Launched Weapons* (London: Issue 19, November 1994).
5 *Washington Post* (11/07/81), p. 16.
6 *Defense Journal* (Karachi: **VIII** (3), March 1982), p. 10.
7 Interview with the Chairman JCSC, General Shamim Alam Khan (Rawalpindi: 13/11/93).
8 *Ibid.*
9 Hasan-Askari Rizvi, *Pakistan and the Geostrategic Environment* (New York: St Martin's, 1993), pp. 92–3.
10 *Aerospace Daily* (28/05/87).
11 *Ibid.* See also, *Defense and Foreign Affairs Daily* (03/04/87).
12 Air Marshal (Retd.) Ayaz Ahmed Khan, 'AWACS and Air Defense of Pakistan'. *Defense Journal* (Karachi: **XIII** (9). 09/10/1987), p. 23.
13 *Jane's Defense Weekly* (09/04/88) p. 658.
14 Interview with the ACAS (Plans) (Rawalpindi: 07/04/94). See also, 'Chengdu J-7/F-7: The Supersonic Sports Plane', *World Power Journal* (7 Autumn/ Winter 1991), p. 134.
15 *Ibid.* pp. 133–4.
16 Interview with Air Chief Marshal (Retd.) Jamal A. Khan (Islamabad: 03/03/94).
17 Interview with Air Marshal (Retd.) Masood Hatif (Islamabad: 04/03/94).
18 *Ibid.*
19 Interview with Air Chief Marshal (Retd.) Abbas Khattak (Rawalpindi: 16/04/94).
20 Interview with Air Marshal (Retd.) Yusuf Khan (Peshawar: 28/03/94).
21 Interview with Air Commodore (Retd.) Sajjad Hyder (Islamabad: 30/01/94).
22 Congressional Presentation on Security Assistance Programs, FY-1983 (Washington: 1983), p. 37.
23 Interview with Lt General (Retd.) Raja Mohammad Iqbal (Rawalpindi: 28/02/94).
24 Interview with the former Army chief, General (Retd.) Gull Hassan Khan (Rawalpindi: 07/11/93). This view was also subscribed to by serving army personnel from the armoured corps during an interview held on 12/12/93.
25 Mary Kaldor, *The Baroque Arsenal* (London: Andre Deutsch, 1982), p. 164.
26 *International Herald Tribune* (13/07/87).

27 Interview with Director R&D, IOP, Iqbal Khan (Rawalpindi: 30/02/94).
28 Interview with Air Commodore (Retd.) Sajjad Hyder (Islamabad: 04/01/94).
29 Interview with the Chief Scientist DESTO, Saleem Mehmood (Rawalpindi: 03/03/94).
30 Interview with Air Commodore (Retd.) Sajjad Hyder (Islamabad: 04/01/94) and Lt General Talat Masood (Islamabad: 02/11/93).
31 Interview with Air Director-General ISPR, Maj. General Salimullah (Rawalpindi: 06/02/98).
32 Mark Montgomery, (US Navy), 'The US–Pak Connection', *Proceedings* (07/89) p. 68.
33 Hasan-Askari Rizvi, *op. cit.*, p. 101. See also, Mark Montgomery, *op. cit.*, p. 70.
34 Interview with Air Commodore (Retd.) Sajjad Hyder (Islamabad: 30/01/94).
35 Interview with Rear Admiral (Retd.) Shamoun Alam Khan (Karachi: 02/12/98). He was the only officer to have put his comments opposing the transfer on file although other officers expressed similar views. Interviews with Commodore (Retd.) M.J.Z. Malik (Islamabad: 08/12/98), Air Commodore (Retd.) G.K. Green (Islamabad: 09/12/98), and Rear Admiral (Retd.) I.H. Naqvi (Islamabad: 10/12/98).
36 Interview with Rear Admiral (Retd.) I.H. Naqvi (Islamabad: October 1998).
37 Interview with Admiral Iftikhar Sirohey (Islamabad: 27/02/94).

8 Military Buildup Decisions, 1990–99

1 Interview with former COMPAK Rear Admiral (Retd) Shamoun Alam (Karachi: 03/05/94).
2 Interview with Admiral Abdul Aziz Mirza (Islamabad: 17/02/98).
3 Interview with naval officers (Islamabad: 01/12/98, 03/12/98 and 04/12/98).
4 *The News* (Lahore: 19/03/93).
5 Interview with Air Marshal Abbas Khattak (Rawalpindi: 16/04/94).
6 *The Muslim* (Islamabad: 07/11/93).
7 *Ibid.*
8 Interview with Additional Secretary, Tariq Fatimi (Islamabad: 17/02/98).
9 Interview with Vice Chief of Air Staff, Air Marshal Aliudin (Rawalpindi: 17/02/98).
10 Interview with Additional Secretary, Tariq Fatimi (Islamabad: 17/02/98).
11 Interview with chief of ISPR, Maj. General Salimullah (Rawalpindi: 28/02/98).
12 Interview with Vice Chief of Air Staff, Air Marshal Aliudin (Rawalpindi: 17/02/98).
13 *Jung* (London: 16/06/95).
14 *The News* (Lahore: 29/09/92).
15 Interview with Additional Secretary Defence Production Division, Air Marshal Qalbe-Abbas Zaidi (Rawalpindi: 31/03/94). See also PACs Brochure on K-8, and *Jane's All the World's Aircraft, 1992–93* (London: Jane's, 1993), p. 128.
16 David Faulkener, *Strategic Alliances. Co-operating to Compete* (London: McGraw-Hill, 1995), pp. 14, 47, 64.

17   Ron Mathews, 'Country Survey IV: Pakistan', *Defence and Peace Economics, V* (York: 1994), p. 325.
18   Interview with Air Marshal (Retd.) Masood Hatif (Islamabad: 04/03/94).
19   Interview with ACAS (Plans) (Rawalpindi: 31/07/97).
20   Interview with Maj. General Salimullah (Rawalpindi: 28/02/98).
21   Interview with Air Vice Marshal Shahid Hamid (Wah: 02/10/98).
22   Interview with the Army's Chief of General Staff, Lt General Farrukh Khan (Rawalpindi: 29/03/94).
23   Interview with Lt General (Retd.) Raja Muhammad Iqbal (Rawalpindi: 28/02/94).
24   Interview with Director General HIT (Taxila: 10/02/98).
25   Interview with Mushahid Hussain (Islamabad: 30/01/94).

9   Mutually Assured Deterrence: the Nuclear Option

1   James E. Dougherty, 'Proliferation in Asia' *Orbis* (Fall 1975–Special Issue), p. 926.
2   Discussion was held in Peshawar (09/05/99).
3   Shahid-Ur Rehman, *Long Road to Chagai* (Islamabad: Print Wise, 1999), pp. 82–5.
4   Thomas C. Schelling, *The Strategy of Conflict* (14th edition) (Massachusetts: Harvard University, 1994), pp. 207–29.
5   Raju G.C. Thomas, *South Asian Security in the 1990s* (London: Adelphi Paper 278, July 1993), p. 68.
6   *Ibid.*
7   Discussion with Admiral (Retd.) L. Ramdas (Ahugalla: 23/09/99).
8   Interview with Special Assistant to Strobe Talbot, Mat Daily (Washington: 08/04/99).
9   Zahid Malik, 'Pakistan's Atomic Programme and General Zia-ul-Haq', in *Shaheed.ul.Islam: Muhammad Zia Ul Haq* (London: Indus Thames, 1990), pp. 80–1.
10 Discussion with Indian defense analyst, Sujit Dutta (Albuquerque: 19/04/99).
11 Kausar Niazi, *Last Days of Premier Bhutto* (Lahore: Jung, 1991), p. 85.
12 Stephen M. Meyer. *The Dynamics of Nuclear Proliferation* (Chicago, 1986), p. 63.
13 David Cortright and Amitabh Mattoo, 'Indian Public Opinion and Nuclear Weapons Policy', in David Cortright and Amitabh Mattoo (eds), *Public Opinion and Nuclear Options* (Notre Dame: University of Notre Dame, 1996), pp. 3–22.
14 Interview with Begum Abida Hussain (Islamabad: 22/03/94).
15 Muhammad Aslam, *Dr. A.Q. Khan and Pakistan's Nuclear Programme* (Rawalpindi: Diplomat, 1989), p. 58.
16 *Ibid.*, p. 11.

10 Looking Ahead

1 *Economic Survey, 1998–99* (Islamabad: Government of Pakistan, Finance Division, Economic Advisor's Wing), pp. 5, 91.
2 Shahid Javed Burki. *Pakistan. Fifty Years of Nationhood* (Boulder: Westview Press, 1999), p. 103.
3 *Economic Survey, 1998–99* (Islamabad: Government of Pakistan, Finance Division, Economic Advisor's Wing), p.2.
4 Agha Shahi, Abdul Sattar, and Zulfiqar Ali Khan. 'Securing Nuclear Peace', *The News*. (Lahore: 05/10/99).

Bibliography

Interviews

Aftab, Moen (Rawalpindi: 08/02/94). Financial advisor (Defence Production Division) in the Ministry of Defence, Rawalpindi.

Afzal, Maj. General (Retd) Muhammad (Rawalpindi: 21/02/94). Director General Heavy Industries, Taxila from 1986–88.

Ahmed, Maj. General (Retd) Naseer (Rawalpindi: 30/11/93) Director-General (Defence Procurement) at the time of the negotiations for Type-21 and Type-23 frigates.

Akhtar, Rear Admiral K.M. (Islamabad: 10/11/93 and Rawalpindi: 27/11/93). Director-General (Defence Procurement) from 1993–95.

Alam, Aroosa (Islamabad: 28/11/93). An Islamabad-based journalist, specializing in defence issues.

Alam, Aslam (Rawalpindi: 01/03/94). Retired in 1992 from a senior position in the Central Inspectorate of Armament, Rawalpindi.

Ali, Lt General (Retd) Ahmed (Rawalpindi: 08/02/94). Served as Additional Secretary Defence Production Division until 1999.

Ali, Shamshaad (Lahore: 04/04/94). A prominent defence subcontractors and member of the executive committee on defense engineering of the Lahore Chamber of Commerce and Industry. The committee was formed in the early 1990s to examine prospects of cooperation with the defence industry.

Anwar, Commander (Karachi: 07/03/94). Commander PNS *Tariq*.

Anwar, Air Vice Marshal Saeed (Rawalpindi: 29/07/97) Director-General, PAC, Kamra.

Arif, Lt General (Retd) K.M. (Rawalpindi: 14/11/93). Retired at the end of 1980s, he was the CGS in 1987–88 at the time of the Indian military exercise, 'Brasstacks'.

Assistant Chief of Air Staff (Plans) (Rawalpindi: 07/04/94).

Azeem, Lt General (Retd) Ijaz (Islamabad: 09/12/93). Served as Pakistan's ambassador to the US during Zia's rule. Also related to General Zia.

Azeem, Air Marshal (Retd) Waqar (Islamabad: 26/04/94). Retired in 1987 after serving in various senior positions in the PAF.

Aziz, Sirtaj (Islamabad: 19/10/97). Federal Finance Minister in Nawaz Sharif's government.

Balmont, Henri (Islamabad: 30/01/94 and 01/02/94). Represented the French Manufacturer Thomson-CSF in Pakistan.

Baqir, Brig. Muhammad (Rawalpindi: 08/02/94). Director Munitions Production (Army).

Bashir, Brig (Retd) (Islamabad: 01/03/94). Head of the Institute of Regional Studies, Islamabad from 1993–97

Bashir, Maj. General Khalid (Rawalpindi: 01/02/94). Director-General ISPR.

Beg, General (Retd) Mirza Aslam (Rawalpindi: 15/05/94). Pakistan's former Army chief. Retired in the early 1990s.

Begg, John R.W. (Karachi: 02/05/94). General Manager (Engineering, Research and Development), Trans-Mobil Limited (TML).

Bhatti, M.A. (Islamabad: 29/01/94). Pakistan's former Ambassador to China, currently working as an expert on Chinese affairs.

Bukhari, Brig. Zulfiqar Ali (Rawalpindi: 06/11/93). Director (Works and Civil Engineering Department) in the Army.

Burki, Maj. General (Retd) Farhat Ali (Rawalpindi: 22/02/94). Director-General, HIT from 1988–92. Prior to that he headed MVRDE and Army Workshop 502.

Cheema, Pervaiz Iqbal (Islamabad: 14/02/94). An expert on defence issues, he was the head of the Department of International Relations, Quaid-i-Azam University, Islamabad.

Chishti, Lt General (Retd) Faiz Ali (Rawalpindi: 08/02/94). He was a member of General Zia's inner policy-making circle until his retirement in 1980.

Cohen, Prof. Stephen P. (Nathiagali: 16/05/94). A South Asia expert, he worked in the US State Department in the 1980s.

Crepin, Bernard (Islamabad: 17/04/94). Head of the French organization 'Office General deell'Air' in Islamabad. This organization represents most of the major French manufacturers in Pakistan.

Daily, Mat (Washington, DC: 08/04/99) Special assistant to Strobe Talbot.

Delassaux, Jean-Marc (Islamabad: 10/02/94). *Chef de Programme* for French manufacturer 'Sagem'.

Deshankar, Prof. Giri (Nathiagali: 14/05/94). Indian expert on China.

Dilawar, Air Marshal Muhammad (Retd) (Kamra: 22/12/93). Director-General PAC, Kamra.

Din, Air Marshal Aliu (Rawalpindi: 17/02/98). Vice Chief of Air Staff.

Din, Lt General (Retd) Kamal Matinnu (Rawalpindi: 23/02/94). One of the first officers to have worked at the JSHQ.

Durrani, Maj. General Mahmud Ali (Wah: 10/12/93 and 12/02/94). Chairman POFs, Wah.

Ezdi, Asif (Islamabad: 05/03/94). Director-General (Europe and CIS), Ministry of Foreign Affairs, Islamabad.

Fatami, Tariq (Islamabad: 17/02/98). Additional Secretary, Ministry of Foreign Affairs. He was later appointed as Pakistan's permanent representative to the UN to be withdrawn soon after Sharif's removal from power.

Gauhar, Altaf (Islamabad: 10/11/93). Member of Field Marshal Ayub Khan's cabinet and one of his most trusted men.

Gull, Lt General (Retd) Hameed (Rawalpindi: 30/01/94 and 15/05/94). Corps Commander at the time of 'Brasstacks', he was retired from the position of the chief of ISI in 1989 by former Prime Minister, Benazir Bhutto.

Hamid, Maj. General Shahid (Taxila: 02/10/98) Director-General HIT, Taxila.

Haq, Dr Mehboobul (Islamabad: May 1997). Economist.

Haq, Dr M. Ikramul (Lahore: 03/04/94). Director of MIRDC, Lahore.

Hatif, Air Marshal (Retd) Masood (Islamabad: 04/03/94). Served in a senior position in the Air Force until his retirement in 1993.

Hussain, Begum Abida (Islamabad: 22/03/94). A prominent politician also served as Pakistan's Ambassador to the US from 1990–93.

Hussain, Commodore Ejaz (Rawalpindi: 27/11/93). Currently a Rear Admiral serving as the DCNS (Maintenance).

Hussain, Mushahid (Islamabad: 30/01/94). Prominent journalist and was the press advisor to Prime Minister Nawaz Sharif from 1991–93 and information minister from 1997–99.

Hussain, Brig (Retd) Noor (Rawalpindi: 09/11/93). Head of the Institute of Strategic Studies, Islamabad until the mid–1980s.

Hyder, Air Commodore (Retd) Sajjad (Islamabad: 04/01/94 and 30/01/94). Retired from the Air Force in the mid–1980s.

Iqbal, Lt General (Retd) Raja Muhammad (Rawalpindi: 28/02/94). Additional Secretary, Defence Division, 1990–92.

Iqtidar, Vice Admiral Muhammad (Karachi: 05/05/94). Head of Karachi Shipyards.

Jalal-u-Din, Wing Commander (Kamra: 22/12/93). Chief Engineer (Facilities & Development), PAC, Kamra.

Kakakhail, Shafqat (Islamabad: 15/05/94). Director-General (India Desk), Ministry of Foreign Affairs, Islamabad.

Khalfan, Abdul Malik (Islamabad: 26/03/94). Vice President of Alsons Industries (Pvt.) Ltd., which is a subcontractor for POF, and an arms broker for MOD.

Khan, Dr Abdul Qadeer (Islamabad: 28/02/94). Head of the Kahuta or Khan Research Laboratories. He is the man reputed to have established Pakistan's nuclear program.

Khan, Amanullah (Rawalpindi: 17/04/94). Head of the Jammu and Kashmir Liberation Front (JKLF).

Khan, Amanullah (Islamabad: 20/04/94). Head of the disbursement section of the World Bank in Pakistan.

Khan, Lt General Farrukh (Rawalpindi: 29/03/94). Retired as CGS of Pakistan Army in 1995.

Khan, Gul (Lahore: 16/10/95). One of the important small arms sellers in Lahore.

Khan, General (Retd) Gul Hassan (Rawalpindi: 07/11/93). He retired as Army Chief in 1972.

Khan, Air Marshal (Retd) I.A. (Lahore: 03/04/94). Retired from the Air Force in the early 1990s after having served in various senior positions.

Khan, Iqbal (Rawalpindi: 30/02/94). Director R&D at the Institute of Optronics, Rawalpindi.

Khan, Air Chief Marshal (Retd) Jamal A. (Islamabad: 03/03/94). Pakistan's Air Chief from 1985–88.

Khan, Lt General (Retd) Muhammad Iqbal (Rawalpindi: 13/02/94). Retired in 1984 after having served as the Chairman JCSC.

Khan, Lt General (Retd) Muhammad Rahim (Rawalpindi: 12/02/94). Served as Secretary-General (Defence) until 1985.

Khan, Munir Ahmed (Islamabad: 10/02/94). Chairman Pakistan Atomic Energy Commission until the mid-1980s.

Khan, Admiral (Retd) Obaidullah (Islamabad: 26/03/94 and 18/05/94). Deputy Chief of Naval Staff (Operations) 1994–96.

Khan, Lt General (Retd) Rahim-u-Din (Rawalpindi: 07/02/94). Retired as Chairman JCSC in 1986.

Khan, General (Retd) Shamim Alam (Rawalpindi: 13/11/93 and 21/03/94). Served as Chairman JCSC until 1995.

Khan, Rear Admiral (Retd) Shamoun Alam (Karachi: 03/05/94). Commander Pakistan Fleet (COMPAK), 1993–95.

Khan, Shehryar (Islamabad: 14/05/94). Served as Pakistan's Foreign Secretary until his retirement in 1994.

Khan, Air Chief Marshal (Retd) Zulfiqar Ali (Islamabad: 09/02/94). Retired as Air Chief in 1978.

Khan, Air Marshal (Retd) Yusuf (Peshawar: 28/03/94). Retired at the end of the 1980s after having served in several important positions including chief of maintenance and Director-General, PAC, Kamra.

Khattak, Air Chief Marshal (Retd) Abbas (Rawalpindi: 16/04/94). Retired as Air Chief in 1997. He was the Deputy Chief of Air Staff (Operations) in 1994.

Malik, Lt General Jamsheed (Rawalpindi: interviewed twice; first as VCGS on 02/12/93 and the second time as Additional Secretary (Defence Division) on 10/02/94).

Martin, Kenneth (Ohio, USA: 15/09/94). Senior director at DISAM.

Masood, Maj. General (Retd) Agha (Rawalpindi: 08/02/94). DGP (Army) and known for establishing the Army's Air Defence Command.

Masood, Lt General (Retd) Talat (Islamabad: 02/11/93, 08/11/93 and 14/02/94). He has the privilege of being Chairman POF and Secretary of Defence (Production Division). He retired from the position as Secretary in 1990.

Mazari, Dr Shireen Islamabad, 24/12/93. Head of the department of Defence and Strategic Studies at Quaid-i-Azam University, Islamabad until 1993. She is held as an expert on defence affairs.

Mehmood Col. (Retd) Ejaz (Islamabad: 18/04/94). Representative in Pakistan for the French manufacturer 'Sofma.'

Mehmood, Saleem (Rawalpindi: 03/03/94). Chief Scientest at DESTO.

Mehsood, Lt General (Retd) Alam Jan (Rawalpindi: 22/02/94). Corps Commander (Lahore) 1989–90.

Mir, Rear Admiral Khalid M. (Karachi: 05/05/94). Commander Karachi until made Chairman Pakistan National Shipping Corporation (PNSC) in 1996.

Mirza, Admiral Abdul Aziz (Islamabad: 17/02/98) Currently serving as the Naval Chief, he was the Vice Chief of Naval Staff at the time of interview.

Musharaf, Maj. General Pervaiz (Rawalpindi: 29/03/94). Director General (Military Operations) at the time interview. Currently the Army Chief and Chief Executive of the Islamic Republic of Pakistan.

Naik, Niaz A. (Islamabad: 23/02/94). Pakistan's former Foreign Secretary.

Najmi, Air Vice Marshal Nafees Ahmed (Rawalpindi: 30/03/94). Held several important positions in the PAF. Head of the Shaheen Foundation until 1999.

Naqvi, Air Commodore Muhammad (Kamra: 22/12/93). Head of the radar construction facility, KARF at PAC.

Naqvi, Brig. Dilbar H. (Rawalpindi: 21/03/94). Served as Director (Intelligence) at JSHQ in 1994.

Naqvi, Lt General Tanveer (Karachi: 26/01/95). Currently serving as the Chairman National Reconstruction Bureau, he was the Director-General, HIT in 1994.

Naqvi, Rear Admiral Tauqir (Rawalpindi: 08/03/94). Served as Additional Secretary (Defence Division), Ministry of Defence.

Nawab, Maj. General (Retd) Ali (Islamabad: 22/02/94). Retired in the early 1970s as the first head of POF.

Nord, Roger H. (Washington, DC: 07/12/97) Division Chief (Policy Development and Review Department) at the IMF.

Paracha, Riaz (Islamabad: 01/12/93). Career diplomat and former Pakistani ambassador.

Parker, Col. Ted (Islamabad: 01/12/93). Representative of the American Security Assistance Organization working in US consulate in Islamabad.

Qadir, Ghulam (Islamabad: 20/04/94). Senior economist working with the World Bank in Pakistan.

Qadir, Rear Admiral (Retd) I.A. (Karachi: 03/05/94). Retired from the Navy in the early 1980s after having served in various senior positions.

Qadir, Rana Abdul (Lahore: 02/04/94). Owner of Atta group of industries – an important defence subcontractor.

Qamaruzaman, Maj. General (Retd) Sabih (Karachi: 04/05/94). Served as Chairman POF until the end of the 1980s when he was made Chairman of the Pakistan Steel Mills, Karachi.

Rana, Farooq (Islamabad: 28/02/94). Served as Director General (East Asia and Pacific Desk), Ministry of Foreign Affairs.

Representative FIAR (Islamabad: 11/03/94). Represented the Italian manufacturer FIAR.

Riaz, Lt General (Retd) Muhammad (Lahore: 02/02/94). Retired in January 1994 as Additional Secretary (Defence Production Division).

Safdar, Lt General (Retd) Muhammad (Lahore: 29/04/94). Appointed in November 1999 as Governor for the largest province of Punjab.

Safdar, Maj. General Muhammad (Lahore: 03/04/94). Served as the GOC (Kashmir and Northern Areas) in 1989–90.

Safi, Ghulam Muhammad (Rawalpindi: 17/04/94). Head of military wing of *Hizb-ul-Mujahideen.*

Salaam, Abid (Wah: 02/05/94). Served as the Controller Factories Accounts, POF, Wah.

Saraf, Prof. Ashraf (Rawalpindi: 16/04/94). Head of the political wing of *Hizb-ul-Mujahideen.*

Sarwar, Brig. Ghulam (Rawalpindi: 02/03/94). Director, MVRDE.

Sarwar, Brig Muhammad (Rawalpindi: 26/02/94). Director, ARDE.

Seth, Hyder (Lahore: 23/09/95). Owner of Buksh Ilahi and Company – one of the largest small arms sellers in Lahore.

Shah, Brig. Hamayun (Islamabad: 09/03/94). Served as the head of Margalla Electronics in 1994.

Shah, Maj. General (Retd) S.H. (Rawalpindi: 20/02/94). The first Director-General of HIT.

Shahi, Agha (Islamabad: 12/11/93 and 29/01/94). Appointed as Foreign Minister (1987–88), he is a career diplomat.

Shamim, Air Chief Marshal (Retd) Anwar (Islamabad: 28/02/94). Served as the Air Chief in the early 1980s.

Sharif, Lt General (Retd) Muhammad (Rawalpindi: 27/02/94). Served as the first Chairman JCSC.

Sheikh, Commodore Mushtaq (Karachi: 07/05/94). General Manager Naval Dockyards, Karachi.

Shinji, Maruyama (Islamabad: 24/04/94). Economic Attaché, Embassy of Japan, Islamabad, Pakistan.
Siddiqui, Saleem (Islamabad: 24/04/94). Director General Audit (Defence Purchases), Department of the Auditor-General of Pakistan.
Singh, Air Commodore (Retd) Jasjit (London: 15/02/95). Director, Institute of Defence Studies and Analysis, New Delhi.
Sirohey, Admiral (Retd) Iftikhar (Islamabad: 27/02/94). Pakistan's Naval Chief until 1988, he was later appointed Chairman JCSC.
Snoek, Henry (Islamabad: 27/03/94). Head of the IMF mission in Pakistan.
Tasneem, Vice Admiral (Retd) Ahmed (Karachi: 07/05/94). Currently the Chairman, Karachi Shipyards and Engineering Works (KSEW).
Ullah, Maj. General Saleem (Rawalpindi: 28/02/98). Director-General, ISPR. He was later posted as the head of Pakistan Rangers, Punjab.
Utra, Maj. General (Retd) Atta Muhammad (Rawalpindi: 21/02/94). Served as Director-General, HIT from 1982–85.
Zaidi, Ijlal Hyder (Islamabad: 27/02/94). Defence Secretary until the early 1990s.
Zaidi, Rear Admiral (Retd) Ghayoor Abbas (Karachi: 03/05/94). Retired from the Navy in 1983.
Zaidi, Air Vice Marshal (Retd) Qalbe Abbas (Rawalpindi: 31/03/94). Served as Additional Secretary (Defence Production Division), Ministry of Defence.

Official documents

Annual Compilation of Appropriation Accounts. 1980–92. Islamabad: Ministry of Finance.
Audit Report Vol. IV Defence Services, 1988–93. Islamabad: Auditor-General of Pakistan, 1993.
Congressional Presentation on Security Assistance Programs, FY-1983. Washington D.C: 1983.
Congressional Record – Senate. Washington, D.C.
Consolidated Manufacturing Accounts of POF 1991–92. Wah: Office of the Controller Factories Accounts, 1992.
Constitution of the Islamic Republic of Pakistan, 1973. Islamabad: Government of Pakistan.
Economic Survey, 1992–93. Islamabad: Government of Pakistan, Finance Division, Economic Advisor's Wing, 1993.
Foreign Military Sales and its Effects on US Economy. Staff Working Paper by the Congressional Budget Office. Washington, D.C.: 23 July, 1976.
GAO Report on 'Tank Co-production Raised Costs and may not meet many Program Goals', presented to the Chairman, Subcommittee on Foreign Operations, Export Financing and Related Programs, Committee on Appropriations, House of Representatives, US Congress, July 1993.
Gulf Crisis, 1990. Rawalpindi: GHQ Publication. (For Reference see *The Times*, London: 14/01/92).
Pakistan's White Paper on Higher Defence Organisation. New Delhi: Strategic Digest, 1976.

US Department of Defense Financial Management Regulation, DoD 7000. 14-R. Vol. 15. Washington, D.C., 1993.

Unpublished theses

Cheema, Zafar Iqbal. 'India's Nuclear Strategy, 1947–91', PhD thesis: Department of War Studies, King's College, University of London, London, August 1991.
Greenlee, Arthur and O'Niell, Michael D. 'Peace Gate: A Case Study of F-16 FMS Management'. MSc thesis: Air Force Institute of Technology, Wright-Patterson Air Force Base, Ohio, September 1984.
Henry, Stephen A. 'The Economics of FMS: An Analysis of the Impact of FMS Policy Change'. MA thesis: Faculty of the School of Engineering, Air Force Institute of Technology, Wright-Patterson Air Force Base, Ohio, September 1978.
Hussain, Vajahat. 'Brass Tacks: A Study of Crisis Management'. MA thesis: Department of Defence and Strategic Studies, Quaid-i-Azam University, Islamabad, 1990.
Kremer, Deborah and Sain, Bill. 'Offsets in Weapon Systems Sales: A Case Study of the Korean Fighter Program'. MSc thesis: Air Force Institute of Technology, Wright-Patterson Air Force Base, Ohio, September 1992.

Books

Arms transfers

Anthony, Ian. *The Arms Trade and Medium Powers. Case Studies of India and Pakistan, 1947–1990*. London: Harvester Wheatsheaf, 1992.
Brzoska, Michael and Ohlson, Thomas. *Arms Transfers to the Third World, 1971–85*. Oxford: Oxford, 1985.
Frank, Lewis A. *The Arms Trade in International Relations*. New York: Pergamon, 1981.
Harkavy, Robert E. *Bases Abroad. The Global Foreign Military Presence*. Oxford: Oxford, 1989.
Joshua, Wynfred and Gibert, Stephen P. *Arms for the Third World: Soviet Military Aid Diplomacy*. Baltimore: Johns Hopkins, 1969.
Kaldor, Mary. *The Baroque Arsenal*. London: Andre Deutsch, 1982.
Kaldor, Mary and Eide, Asbjorn (eds). *The World Military Order. The Impact of Military Technology on the Third World*. London: Macmillan, 1979.
Kolodziej, Edward A. *Making and Marketing Arms: The French Experience and its Implications for the International System*. New Jersey: Princeton University Press, 1987.
Kolodziej, Edward A. and Pearson, Fredric S. *The Political Economy of Making and Marketing Arms: A Test for the Systemic Imperatives of Order and Welfare*. Illinois: University of Illinois, Urbana-Champaign.
Nincic, Miroslav. *The Arms Race. The Political Economy of Military Growth*. New York: Praeger, 1982.
Pierre, Andrew J. *The Global Politics of Arms Sales*. New Jersey: Princeton University Press, 1982.

Sampson, Anthony. *The Arms Bazaar* (First edition). London: Coronet Books, 1977.

Schelling, Thomas C. *The Strategy of Conflict*. Massachusetts: Harvard University Press. 1994.

Sheehan, Michael. *The Arms Race*. Oxford: Martin Robertson, 1983.

Stanley, John and Pearton, Maurice. *The International Trade in Arms*. London: Chatto & Windus, 1972.

The Arms Trade with the Third World. Stockholm: SIPRI, 1971.

Arms production

Badiru, Adedeji Bodunde. *Managing Industrial Development Projects*. New York, 1993.

Brzoska, Michael and Ohlson, Thomas. *Arms Production in the Third World*. Stockholm: SIPRI, 1986.

China Today: Defence Science and Technology. Beijing: National Defence Industry Press, 1993.

Faulkner, David. *Strategic Alliances. Co-operating to Compete*. London: McGraw-Hill, 1995.

Gansler, Jacques S. *The Defense Industry*. Massachusetts: Lexington, 1989.

Hussain, Nazir. *Defence Production in the Muslim World*. Karachi: Royal, 1989.

Ikegami-Anderson, Masako. *The Military-Industrial Complex. The Case Studies of Sweden and Japan*. Aldershot: Dartmouth, 1992.

Katz, James Everett. *Arms Production in Developing Countries*. Massachusetts: Lexington, 1984.

Krause, Keith. *Arms and the State: Patterns of Military Production and Trade*. Cambridge: Cambridge, 1992.

Kuzmak, Arnold M. *Naval Force Level and Modernisation. An Analysis of Shipbuilding Requirements*. Washington, D.C.: Brookings Institution, 1971.

Malecki, Edward J. *Technology and Economic Development*. New York: Longman Scientific & Technical, 1991.

Reiser, Stewart. *The Israeli Arms Industry*. New York: Holmes & Meier, 1989.

Todd, Daniel. *Defence Industries. A Global Perspective*. London: Routledge, 1988.

Decision-making

Allison, Graham T. *Essence of Decision. Explaining the Cuban Missile Crisis*. Boston: Little Brown, 1971.

Almond, Gabriel A. *The American People and Foreign Policy* (Second edition). Connecticut: Greenwood Press Publishers, 1977.

Blechman, Barry M. *The Politics of National Security*. Oxford: Oxford, 1990.

Braybrooke, David. and Lindblom, Charles E. *A Strategy of Decision: Policy Evaluation as a Social Process*. London: Macmillan, 1970.

Fallows, James. *National Defense*. New York: Vintage, 1981.

Halliday, Fred. *The Making of the Second Cold War* (Second edition). London: Verso, 1986.

Halperin, Morton H. *Bureaucratic Politics & Foreign Policy*. Washington, D.C.: Brookings Institution, 1974.

Halperin, Morton H. *National Security Policy Making*. Massachusetts: Lexington, 1975.

Harrison, E.F. *The Managerial Decision-Making Process*. Boston: Houghton Mifflin, 1987.

Hilsman, Roger. *The Politics of Policy Making in Defense and Foreign Affairs*. New York: Columbia University, 1971.

Johnson, Franklyn A. *Defence by Ministry*. London: Duckworth, 1980.

Mclean, Scilla (ed.). *How Nuclear Weapons Decisions are Made*. London: Macmillan, 1986.

Murray, Douglas J. and Viotti, Paul R. (eds). *The Defense Policies of Nations* (Second edition). Baltimore: Johns Hopkins University Press, 1989.

Pfaltzgraff Jr, Robert L. and Davis, Jacquelyn K. (eds). *National Security Decisions. The Participants Speak*. Massachusetts: Lexington, 1990.

Simon, Herbert A. *Administrative Behaviour: A Study of Decision-Making Process in Administrative Organisations*. New York: Macmillan, 1947.

Smith, Perry M, Maj. General (Retd). *Assignment: Pentagon*. Washington, D.C.: Brassey's, 1993.

Defense economics

Deger, Saadet and Sen, Somnath. *Military Expenditure: The Political Economy of International Security*. Oxford: Oxford, 1990.

Hewitt, Daniel P. *Military Expenditure: Econometric Testing of Economic and Political Influences*. IMF Working Paper, May 1991.

Hobkirk, Michael D. *The Politics of Defense Budgeting*. Washington, D.C.: National Defense University Press, 1983.

Kennedy, Gavin. *Defence Economics*. London: Duckworth, 1983.

McKinlay, Robert. *Third World Military Expenditure. Determinants and Implications*. New York: Pinter Publishers, 1989.

Schilling, Warner R., Hammond, Paul Y. and Snyder, Glenn H. *Strategy, Politics and Defence Budgets*. New York: Columbia University Press, 1962.

Schmidt, Christian (ed.). *The Economics of Military Expenditure*. London: Macmillan, 1987.

Nuclear proliferation

Ali, Akhtar. *Pakistan's Nuclear Dilemma*. New Delhi: ABC Publishing House, 1984.

Aronson, Shlomo with Brosh, Oded. *The Politics and Strategy of Nuclear Weapons in the Middle East*. New York: State University of New York Press, 1992.

Aslam, Muhammad. *Dr. A.Q. Khan and Pakistan's Nuclear Programme*. Rawalpindi: Diplomat, 1989.

Barnaby, Frank. *The Invisible Bomb*. London: I.B. Taurus, 1993.

Creveld, Martin Van. *Nuclear Proliferation and the Future of Conflict*. New York: Free Press, 1993.

Gray, Colins. *Weapons Don't Make War*. Kansas: University Press of Kansas, 1993.

Kapur, Ashok. *Pakistan's Nuclear Development*. London: Croom Helm, 1987.

Malik, Zahid. *Dr. A.Q. Khan and the Islamic Bomb*. Islamabad: Hurmat, 1992.

Moshaver, Zeba. *Nuclear Weapons Proliferation in the Indian Subcontinent*. London: Macmillan, 1991.

Niazi, Kauser. *Last Days of Premier Bhutto*. Lahore: Jung, 1991.

Subamanian, R.R. *India, Pakistan, China: Defence and Nuclear Tangle in South Asia.* New Delhi: ABC Publishing House, 1990.

South Asia politics and security

Ahmed, Munir. *Political Role of Intelligence Agencies in Pakistan* (Urdu edition). Lahore: Jahangir Book Depot, 1993.

Burke, S.M. and Ziring, Lawrence. *Pakistan's Foreign Policy: An Historical Analysis.* Karachi: Oxford, 1990.

Cheema, Pervaiz Iqbal. *Pakistan's Defence Policy, 1947–58.* London: Macmillan, 1990.

Cohen, Stephen P. *The Pakistan Army.* Karachi: Oxford, 1984.

Gupta, Shekhar. *India Redefines its Role* (Adelphi Paper 293). Oxford: Oxford, January 1995.

Huang, Prof. Yasheng *China's Economic Development: Implications for its Political and Security Roles* (Adelphi Paper 275). London: IISS, March 1993.

Hussain, Mushahid. *Pakistan's Politics: The Zia Years.* Lahore: Progressive, 1990.

Hussain, Mushahid and Hussain, Akmal *Pakistan. Problems of Governance.* Lahore: Vanguard, 1993.

Isphani, Mehnaz. *Pakistan: Dimensions of Insecurity* (Adelphi Papers 246). London: Winter 1989–90.

Jalal, Ayesha. *Democracy and Authoritarianism in South Asia.* Lahore: Sange Meel, 1995.

James, Sir Morrice. *Pakistan Chronicle.* Karachi: Oxford, 1993.

Kennedy, Gavin. *The Military in the Third World.* New York: Charles Scribner's Sons, 1974.

Khan, Zulfiqur Ali. *Pakistan's Security: The Challenge and the Response.* Lahore: Progressive, 1988.

Kukreja, Veena. *Civil–Military Relations in South Asia.* New Delhi: Sage, 1991.

Lamb, Alastair. *Kashmir: A Disputed Legacy, 1846–1990.* Hertfordshire: Roxford, 1991.

Malik, Hafeez. *Soviet-Pakistan Relations and Post-Soviet Dynamics.* London: Macmillan, 1994.

Musa, Muhammad, General (Retd). *My Version: India–Pakistan War, 1965.* Lahore: Wajidalis, 1983.

Nasr, Seyyed Vali Raza. *The Vanguard of the Islamic Revolution. The Jama'at-i-Islami of Pakistan.* London: I.B. Taurus, 1994.

Noman, Omar. *Pakistan: Political and Economic History Since 1947.* New York: Kegan Paul, 1988.

Rikhye, Ravi. *The Militarisation of Mother India.* New Delhi: PRISM, 1990.

Rikhye, Ravi. *The War that Never Was.* Lahore: Army Education Press, 1989.

Rizvi, Hasan-Askari. *The Military and Politics in Pakistan.* Lahore: Progressive, 1974.

Rizvi, Hasan-Askari. *Pakistan and the Geostrategic Environment.* New York: St Martin's, 1993.

Roy, Oliver. *The Lessons of the Soviet/Afghan War* (Adelphi Paper 259). London: Summer 1991.

Shahi, Agha. *Pakistan's Security and Foreign Policy.* Lahore: Progressive, 1988.

Stoessinger, John G. *Why Nations Go to War* (Fifth edition). New York: St Martin's, 1990.

Thomas, Raju G.C. *South Asian Security in the 1990s* (Adelphi Paper 278). London: July 1993.

Tully, Mark and Jacob, Satish. *Amritsar: Mrs Gandhi's Last Battle.* London: Jonathan Cape, 1985.

Yousaf, Muhammad and Adkin, Mark. *The Bear Trap: Afghanistan's Untold Story.* Lahore: Jung, 1992.

Wirsing, Robert G. *Pakistan's Security under Zia, 1977–1988.* London: Macmillan, 1991.

Wirsing, Robert G. *India, Pakistan and the Kashmir Dispute. On Regional Conflict and its Resolution.* London: Macmillan, 1994.

Weapons procurement

Edmonds, Martin. *International Arms Procurement. New Directions.* New York: Pergamon Press, 1981.

Jones, Rodnet W. and Hildreth, Steven A. *Modern Weapons and Third World Powers.* Boulder: Westview Press, 1984.

Klieman, Aharon and Pedatzur, Reuven. *Rearming Israel: Defence Procurement Through the 1990s.* Boulder: Westview Press, 1991.

'Two-Way Street.' *The Klepsch Report on USA–Europe Arms Procurement.* London: Brassey's, 1979.

Index